™

Outer Space
Directory

The Products, Places, and People Directory™

by

Spencer Kope

1998 Edition
Willow Creek Press of Washington

KOPE'S OUTER SPACE DIRECTORY
The Products, Places, and People Directory™
FIRST EDITION

Cover design by Willow Creek Press of Washington
Kope's logo by Brian Koningisor (Sunshine Graphics, Bremerton, Washington)

Cover Photos:

Moonbeam Meteorites Necklace: David Leviton
Renaissance Telescope: Tele Vue
Mars Global Surveyor Model: Space Craft International
Astronomy Aids: David Chandler Company
Apollo 13 Command Module: Kansas Cosmosphere
STARLAB: Learning Technologies
Adler Planetarium: Adler Planetarium
Beyond the Home Planet: Lynette Cook
Kennedy Space Center: NASA

Publisher's Cataloging-in-Publication
(Provided by Willow Creek Press)

Kope, Spencer.
     Kope's outer space directory: the products, places, and people directory / by Spencer Kope. -- 1998 ed.

     p. cm.
     Includes indexes.
     ISBN: 0-9647183-2-4

     1. Astronomy--Miscellanea--Directories. 2. Outer space--Miscellanea--Directories. I. Title.

QB47.K67 1998   522'.025
                         QBI97-41073

Price: $19.95
Willow Creek Press of Washington
P.O. Box 3730, Silverdale, WA 98383-3730
Phone: (360) 830-5612, E-mail: kope@sincom.com

Table of Contents

Introduction

Welcome to the first edition of *Kope's Outer Space Directory*. A brief explanation of this book, how it was put together, and how to use it is in order. In gathering the information for this book, Willow Creek Press of Washington requested information from astronomy and space-related companies, observatories, planetariums, museums, organizations, associations, and various federal institutions. Unfortunately not all organizations responded to our request.

It is the policy of Willow Creek Press to provide descriptive articles on only those organizations which respond to our request for information. If you don't find a descriptive entry for an organization in the body of the book, check the appendices. There are two appendices which list additional organizations, and two that are NASA-related. If you don't find a descriptive entry for a particular planetarium, observatory, or other place of interest, refer to the back of the *Places of Interest* section for a list of institutions that did not provide information. This data is limited to the name and address of the site.

The *Products & Services* section identifies companies and organizations which offer astronomy and space-related products and services. Below the address of each company is a product list that identifies the types of products and services offered by the company. This list is not meant to be all-inclusive, but instead, it identifies the *primary* products and services offered by each company. Many companies offer a variety of other products not listed in their write-up. This was done because of space constraints. With the extensive inventories of many of the companies in this book, it would take too much space to list each item individually.

The *Places of Interest* and the *Organizations* sections are each followed by a QuickFind Index. These QuickFind Indexes break the *Places of Interest* and *Organizations* sections into an easy to use, state by state listing. The idea was to make it easier to identify museums and other sites by location, particularly if you don't know the name of the site. This way, if you're going to Tennessee on vacation (or business), you can quickly identify those sites in Tennessee that you may not have otherwise known about.

The rest of the book is self-explanatory. If you're having trouble finding something, you can always refer to the Master Index at the back of the book.

Well, that's it. We had a lot of fun putting this book together and hope it leads you to new discoveries in astronomy and space exploration.

Spencer Kope
Editor

*I'd like to say a special **"Thank You"** to all the companies, organizations, and institutions that helped make this book what it is. You've all been a great help. This book is a reflection of your passion and knowledge.*

Products & Services

Adirondack Video Astronomy

35 Stephanie Lane
Queensbury, NY 12804
Phone: (518) 793-9484
Fax: (518) 747-2137
E-mail: 72323.3043@compuserve.com

Product(s): CCDs, Imaging Accessories, Telescope Accessories, Video Systems

Photo: Adirondack Video Astronomy

The ASTROVID 505E B&W CCD Video Camera

Specializing in astronomical electronic imaging, Adirondack Video Astronomy (AVA) manufactures the ASTROVID™ line of CCD video cameras, including the ASTROVID 400, ASTROVID 505E, and ASTROVID 7000. The ASTROVID 400 is a black and white high resolution CCD video camera. At the other end of the spectrum is the ASTROVID 7000, a high resolution (500 lines of horizontal resolution) color CCD video system. AVA systems include the camera, power supply, 25' cable with connectors, T-C adapters, instructions and a one year warranty.

Video astronomy involves the use of high resolution CCD video cameras for real-time observation and recording of astronomical objects and events. AVA offers a number of accessory items for your ASTROVID™ system, such as Snappy, a frame grabbing system that allows you to capture still images from your live or taped video feed.

Adorama, Inc.

42 West 18th Street
New York, NY 10011
Phone: (212) 741-0401
Store: (212) 647-9800
Fax: (212) 727-9613
E-mail: goadorama@aol.com

Product(s): Binoculars, Telescopes, Telescope Accessories

Adorama carries a complete line of telescopes and accessories from companies like Celestron, Tele Vue, Bushnell, Lumicon, Telrad, JMI, and others. In addition, a large selection of astronomical binoculars is available, as well as an extensive line of eyepieces, eyepiece filters, and a line of giant binoculars with 70, 80, and 100mm apertures. Their large telescope display, dedicated staff, and extensive inventory make Adorama an important resource for astronomers. A catalog is available upon request.

Andromeda Software, Inc.

P.O. Box 605
Amherst, NY 14226-0605
Fax: (716) 691-6731
E-mail: androsw@frii.com
Internet: www.frii.com/~androsw/

Product(s): Computer Software

Andromeda Software, Inc. has been serving the needs of the astronomical computing community since 1985. Current offerings include a large selection of astronomy and scientific CD-ROMs for IBM compatible and MacIntosh computers. Also available is a huge collection of astronomy and science shareware for IBM & compatible computers. Astronomy categories include desktop planetariums, astronomical databases, orbital mechanics, satellite prediction, gravitation simulations, and image sets. Science categories include earth science, space exploration, geology, archeology, math, medical, physics and much more. The Andromeda Software, Inc. Home Page (http://www.frii.com/~androsw/) contains complete on-line catalog listings and ordering instructions. As a service to customers, the web site also features detailed CD-ROM reviews with numerous full color screen images to help make the CD-ROM selection process as easy as possible. A free product list is available (specify IBM or Mac).

Apogee Inc.

P.O. Box 136
Union, IL 60180
Phone: (815) 923-1602
Fax: (815) 923-1602
E-mail: AstroSirPlus@il-icom.net
Internet: www.Astronomy-Mall.com

Product(s): Telescope Accessories

Apogee offers a wide selection of optical and mechanical surplus items for the amateur astronomer, including mirrors, achromats, simple lenses, beam splitters, AC clock motors, DC motors, AC & DC gearhead motors, stepper motors, telescope tubing (fiberglass & aluminum), eyepieces, finders, dew chasers, filters, aluminized mylar, clock drive kits, gears, bearings, camera lenses, batteries, Barlow & filter protection tubes, muffin fans, and more. Call or write for a free catalog.

Astro Cards

P.O. Box 35
Natrona Heights, PA 15065
Phone (412) 295-4128
E-Mail: AstroCards@aol.com
Internet: Astronomy-Mall.com

Product(s): Slides, Star Maps, Videos

Astro Cards offers the latest images (on color slides) from the Hubble Space Telescope and a full line of images of the planets and other objects transmitted to Earth by JPL & NASA space probe missions to Mercury, Venus, Mars, Jupiter, Saturn, Uranus, Neptune, the Earth and its Moon. For the telescope user, Astro Card Finder Charts, printed on 3x5 index cards, show how to find thousands of interesting deep-sky objects. Call or write for a free price list.

Astronaut Connection, Inc.

Phone: (610) 941-0398
Fax: (610) 941-6799
E-mail: shurst@dmatrix.com
Internet: www.nauts.com

Service(s): Astronaut Appearances

Astronauts have accomplished and experienced things that inspire the dreams and harness the imagination of millions of people around the world. They are universally trusted, respected and revered. People of all ages and backgrounds have a fascination with these intrepid explorers because they are, in every sense of the word, true American heroes.

Now, your business, organization or school can enjoy a visit by one of these space exploration heroes. Astronaut Connection is proud to be a single contact point for over 35 retired astronauts and cosmonauts who are available for commercial appearances, dynamic presentations, promotional programs, or corporate and product endorsements. In addition, astronauts can participate in staff dinners, sales conferences, golf outings, advertising, or serve as corporate spokesmen, key note speakers, seminar hosts, and more.

When a retired Astronaut appears they bring the excitement of the unknown. Astronaut appearances add a high-profile exciting element to any program and have been proven to deliver unparalleled results. Retired astronauts are available for your event for as little as $1,500 per appearance, with fees that vary depending on the retired astronaut selected.

A sample of available retired NASA Astronauts and Payload Specialist Astronauts includes Dr. Joseph Allen (STS-5, STS-51A), Alan Bean (Apollo 12, Skylab 3), Dr. Guion "Guy" Bluford, Jr. (STS-8, STS-61A), Scott Carpenter (Mercury 7), Gerald Carr (Skylab 3), Robert Cenker (STS-61C), Charles "Pete" Conrad (Gemini 5 & 11, Apollo 12, Skylab 1), Charles Duke (Apollo 16), and Dr. Edgar Mitchell (Apollo 14).

Astronomics

2401 Tee Circle
Suites 105/106
Norman, OK 73069
Phone: (405) 364-0858
Fax: (405) 447-3337
Orders: 1-800-422-7876
E-mail: okastro@aol.com
Internet: www.astronomics.com

Product(s): Binoculars, Books, CCDs, Telescopes, Telescope Accessories

If you're new to astronomy and are not quite sure what type of telescope fits your observing needs, Astronomics has an excellent 8-page brochure called "How to Pick the Right Telescope"

Astronomics has an excellent 8-page brochure called "How to Pick the Right Telescope"

which will help explain the types of scopes and their strong and weak points. This brochure is just one of the ways Astronomics helps prospective customers purchase a telescope they won't be disappointed in. Another way they achieve this is by offering quality scopes, like Meade, Tele Vue, Celestron, and others. A full line of telescope accessories is available, from dew remover systems and telescope computers to drive motors and dust covers.

Astro-Physics, Inc.

11250 Forest Hills Road
Rockford, IL 61115
Phone: (815) 282-1513
Fax: (815) 282-9847

Product(s): Telescopes, Telescope Accessories

Astro-Physics is a professional optical and mechanical production facility dedicated to the production and development of amateur telescopes and accessories. Telescopes currently available include Astro-Physics' 130mm f6 Starfire EDFS, 155mm f7 Starfire EDFS, and 155mm f7 Starfire EDF. A full line of accessories is available, including several equatorial mounts designed and manufactured by Astro-Physics.

Photo: Astro-Physics, Inc.

155mm StarFire EDFS with 2.7" focuser, 600E German Equatorial, hardwood tripod and accessories.

Astrostock

2195 Raleigh Ave.
Costa Mesa, CA. 92627
Phone: (714) 722-7900
Fax: (714) 646 7578
E-mail: JRSanf@aol.com
Internet: www.Astronomy-Mall.com

Product(s): CCDs, Photographs

Astrostock services the media with stock images of astronomical subjects, including the space art of Chris Butler. They lease the use of their transparencies to publishers, authors, CD-ROM publishers, etc. on a one-time basis. Clients have included Time, Discover, Holt, Rinehart and Winston, Simon & Schuster, Strange Universe, and others.

Astrostock also sells astronomical CCD cameras from Pegasus Electronics, and their own line of small, real-time CCD video cameras. Accessories such as CCD flip mirrors, filters, monitors, and others are also offered. Check out their web site for more information.

AstroSystems, Inc.

5348 Ocotillo Court
Johnstown, CO 80534-9322
Phone: (970) 587-5838
Fax: (970) 587-4613
E-mail: astrosys@frii.com
Internet: www.frii.com/~astrosys

Product(s): Charts, Meteorite Jewelry, Telescopes, Telescope Accessories

Those interested in building their own telescope will find the AstroSystems catalog helpful and informative. A variety of kits, components, and accessories are available, including the TeleKit Truss Tube Dobsonian which you assemble. Sizes range from 10" to 30" with prices starting at $950, plus shipping. AstroSystems offers optics from Galaxy Optics, Nova Optics, and Pegasus Optics, telescope covers from Scopecoat, Sky Atlas 2000.0 and other reference material, Telrad sights, meteorite jewelry (such as an etched Gibeon Meteorite Watch), and a variety of other items.

Astrovisuals

6 Lind Street
Strathmore VIC 3041
Australia
Phone: (03) 9379 5753
Phone: 61-3-9379 5753 (international)
Internet: www.ozemail.com.au/~astrovis/

Product(s): Computer Software, Postcards, Posters, Slide Presentations, Videos

Astrovisuals offers a nice assortment of astronomical visual materials, including several multimedia CDs containing images, narration, and music. A sampling of videos includes *Apollo 13: Houston We've Got a Problem* (the official NASA documentary from 1970), *Earthviews* (a selection of over 80 images of the Earth from space), and *Universe*, narrated by William Shatner. Three astronomy slide units are currently available, each with 20 slides. These units are *Hubble*, *Planets*, and *Deep Space* (David Malin, AAO Images).

Photos: Astrovisuals

Posters available from Astrovisuals include Starbirth, The Earth, *and* The Milky Way.

Joe Bergeron

2901 Hall St Apt 2
Endwell, NY 13760
Phone: (607) 786-0754
E-mail: JABergeron@aol.com
Internet: members.aol.com/
jabergeron/index.html

Product(s): Art

Joe Bergeron specializes in space illustration and space-related computer graphics. Samples can be viewed at Mr. Bergeron's web site.

Blue Chromium Records

P.O. Box 50358
Pasadena, CA 91115
Phone: (626) 791-1480
Fax: (626) 794-5751
E-mail: markmercury@compuserve.com

Product(s): Music (space)

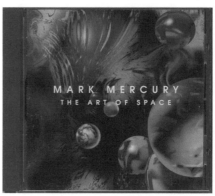

Photo: Blue Chromium Records

Featuring the space music of Mark Mercury, whose compositions can be heard in planetariums across the country, releases by Blue Chromium Records include *The Art of Space*, a musical fantasy that chronicles the thoughts and emotions of a lone space traveler spinning into the cosmos, and *Music of the Domes*, a collection of Mark Mercury's planetarium composi- tions. With impressive credentials that include compositions for film, televi- sion, and commercials, Mark Mercury is one to watch — and listen to.

Brite Sky

5b First Street
Dover, NH 03820
Phone: (603) 743-4083
Fax: (603) 742-5456

Product(s): Optics Cleaning Fluid

Brite Sky offers a new optics cleaning fluid that promises to increase resolution and light transmission by re- storing your optics to their original per- formance level. Brite Sky removes surface residue and residues at the microscopic level to the point of even removing oils from the pores of your optics. For Schmidt Cassegrain owners, cleaning with Brite Sky helps slow dew formation on your collector plate. Not intended for use on surface coat mirrors. Brite Sky is available at most telescope and binocular dealers.

Carina Software

12919 Alcosta Blvd., Suite #7
San Ramon, CA 94583
Phone: (510) 355-1266
Fax: (510) 355-1268
E-mail: support@carinasoft.com
Internet: www.carinasoft.com

Product(s): Computer Software

Carina Software is a premier publisher of astronomy software for the MacIntosh. Their flagship product, *Voyager II, the Dynamic Sky Simulator,* received a 5-star rating when reviewed and is highly regarded for its depth of features and ease of use. A newer program called *SkyGazer* is specifically designed for the novice and the education market. Carina also offers *SkyPilot,* an interface for popular commercial telescopes, making possible a whole new experience in celestial observing. Windows 95 versions are planned for 1998.

Carl Zeiss Optical, Inc.

1015 Commerce Street
Petersburg, VA 23803
Phone: 1-800-338-2984
Fax: (804) 733-4024

Product(s): Binoculars

Zeiss, world renowned for precision optical products, offers a full-line of binoculars through Carl Zeiss Optical, Inc. For a dealer near you or for a free brochure call Carl Zeiss Optical, Inc. on their toll free line, 1-800-338-2984.

Photo: Carl Zeiss Optical

Photo: Carl Zeiss Optical

Carl Zeiss Optical offers a nice selection of binoculars for both terrestrial and astronomical viewing.

Celestis, Inc.

2444 Times Blvd., Suite 260
Houston, TX 77005
Phone: (713) 522-7282
Fax: (713) 522-7380
Information: 1-800-ORBIT 11
E-mail: celestis@iah.com
Internet: www.celestis.com

Service(s): Space Burial

Space-lovers around the world now have the incredible opportunity to have their ashes launched into space onboard a memorial space flight that circles the earth and eventually ends in the most brilliant of tributes—as a magnificent shooting star. This special out-of-this-world service can be purchased when needed, or you can become a Celestis Associate Charter Member now and have the additional opportunity to express yourself by way of a personal message (25 words or less) which is launched into space. Your message will be placed on the next available launch, orbit the earth and then literally become a shooting star as it re-enters the upper atmosphere. After your launch you will receive a photo/certificate of the launch displaying the actual launch, launch date, orbital data and your name and personal message—authenticated by Celestis as a memento of your mission.

When you join, you will also receive a Charter Membership card which serves as evidence of your good standing and will allow you access to the Celestis launch activities. It also serves to provide important information to your family regarding your final wishes in the event of need. Current members will receive a 10% discount on the cost of the Celestis Service if purchasing for themselves for future need or for a loved one for immediate need. This is a savings of $480 off of the price of the Celestis Earthview Service, and may represent even greater savings on future Celestis Voyager missions.

Celestis Associate Members will also receive $10,000 in accidental death insurance protection for the twelve month period of membership. This protection is included in the basic membership and assures that funds will be available for the Celestis Service in the event of accidental loss of life. No salesman will call and there are no premiums that you must pay. This coverage is provided to Members through a master policy held by Celestis, Inc. A detailed explanation of benefits will be provided to you.

To purchase the Celestis Earthview Service, you can contact Celestis directly, or make arrangements through your local funeral home. The cost of the service is $4800 (US$), though, again, Celestis Associate members receive a 10% discount (that's an immediate savings of $480).

Celestis, Inc. also has a non-profit affiliate, The Celestis Foundation, which focuses on nurturing entrepreneurial space enterprises, supporting organizations that educate our children and the general public about space, and contributing to charities that create a positive future on Earth. The internationally recognized experts and leaders serving on the advisory board recommend worthy projects for the annual Celestis Foundation grants, first awarded in 1995.

Your contribution to the Celestis Foundation ensures a continuing source of support for the people and projects that will accelerate the opening of the space frontier and the preservation of Earth.

Clear Skies

10300 S.W. 4th Avenue
Portland, OR 97219
Phone: (503) 452-9634

Service(s): Portable Planetarium

Educators looking for a way to offer field trips without the hassle and expense of packing students onto a bus and heading across town will be delighted to hear about Clear Skies.

This company's portable planetarium is a 16-foot fan inflated dome that seats approximately 30 people and is capable of doing what larger fixed planetariums do. Last year Clear Skies brought the magic of their portable planetarium to over 22,000 people, doing programs for the clubs, libraries, resorts, schools, and more. Sounds like a fun addition to any star party, especially if the weather forecast is questionable.

Cobblestone Publishing Company

Simon & Schuster Education Group
7 School Street
Peterborough, NH 03458-1454
Phone: (603) 924-7209
Orders: 1-800-821-0115
Internet: www.cobblestonepub.com

Product(s): Magazines

Cobblestone is the publisher of the award-winning astronomy and space science magazine, *Odyssey*. Written to stimulate young readers to explore the wonders of space and understand the importance of science, every 49-page issue of Odyssey features science articles, interviews with scientists, photographs and original illustrations, classroom and home activities, and more. Odyssey is published nine times a year and classroom subscriptions are available.

Lynette R. Cook

1548 45th Avenue
San Francisco, CA 94122
Phone/Fax: (415) 564-2757
E-mail: lrcook@sirius.com

Product(s): Art

Lynette Cook is a talented freelance scientific illustrator who specializes in astronomical imagery. Her subject matter includes extra solar planets, life in space, SETI, galaxies, comets, and more. Imagery is available for non-exclusive use and originals are for sale. Lynette's work has appeared in Astronomy, Final Frontier, and Earth magazines; Time-Life Books; documentaries produced by BBC Television, The Learning Channel, and Swedish Television; and commercial products by The Nature Company (we told you she was talented).

Art: Lynette Cook

Cosmic Resources

Bill Schultz
5659 North Glen
Cincinnati, OH 45248
Phone: (513) 574-2226

Service(s): Consulting, Lectures, Stargaze Programs

Bill Schultz is the Education Director for the Cincinnati Astronomical Society (CAS) and provides a variety of services to educators and others with an interest in astronomy. Mr. Schultz is available for public lectures, stargaze programs, tips on buying a telescope, and astronomical advice. He has appeared on television and in newspapers and is named in several resource books.

D&G Optical

6490 Lemon Street
East Petersburg, PA 17520
Phone: (717) 560-1519
Fax: (717) 581-9015
E-mail: dgoptical@aol.com

Product(s): Telescopes, Telescope Accessories

Priding itself on the quality of its lenses and mirrors, D&G Optical insures that each lens is individually hand-figured to 1/20 wave surface accuracy and star tested prior to shipment. All lenses are mounted in push-pull adjustable cells and are available in 5, 6, 8, 10, and 12-inch sizes. D&G also offers achromat tube assemblies, Cassegrain optics and tube assemblies, focusers, mounting rings, finders, mounts, and more.

David Chandler Company

P.O. Box 999
Springville, CA 93265
Phone: (209) 539-0900
Fax: (209) 539-7033
Orders: 1-800-516-9756
E-mail: dschandler@frumble.claremont.edu
Internet: www.csz.com/dschandler

Product(s): Astronomy Kits, Books, Computer Software, Planispheres

"a variety of useful aids for the amateur astronomer"

The David Chandler Company offers a variety of useful aids for the amateur astronomer, including the award winning guide *Exploring the Night Sky with Binoculars*, the *Sky Atlas For Small Telescopes*, the *First Light Astronomy Kit*, the *Star Light Binocular Astronomy Kit*, and *The Night Sky* two-sided planisphere.

Also available is *Deep Space Version 5*, a full-featured observing guide that generates publication quality star maps, serves as an almanac, an ephemeris and finder chart generator, an annotated deep sky reference, an observing log utility, a telescope controller, and more. The *Deep Space* CD ROM contains the NASA Skymap database of 250,000 stars, the Hubble Guide Star Catalog of 18 million objects, the Saguaro database of 10,000 deep sky objects, the JPL DASTCOM database of 10,000 asteroids and comets, and the CBAT comet database of 1100 comets.

Photo: Earth & Sky

Earth & Sky

P.O. Box 2203
Austin, TX 78768
Phone: (512) 480-8773
Fax: (512) 477-4474
E-mail: info@earthsky.com
Internet: www.earthsky.com

Service(s): Radio Program

Deborah Byrd and Joel Block have been presenting the Earth & Sky science series for nearly two decades. The program is heard on more than 650 radio stations in the U.S., and many more around the world. Topics include Earth science, astronomy, environmental science, and other sciences. Listener questions are encouraged. Contact Earth & Sky to play the program locally, ask questions or sponsor programs.

Joel Block and Deborah Byrd, hosts of Earth & Sky

Edward R. Byers Company

29001 West Highway 58
Barstow, CA 92311
Phone: (760) 256-2377
Fax: (760) 256-9599

Product(s): Telescope Accessories

Edward R. Byers offers precision worm gears and C-14 retrofits. In 1998, they will begin offering mounts, Cam-Traks, and focusers. A catalog is available for $2 ($5 foreign).

Energems Unlimited

83-5782 Napoopoo Rd.
Captain Cook, HI 96704
Phone: (808) 328-9217
Fax: (808) 328-8058
E-mail: ranger2@aloha.net

authenticated meteorite specimens

Product(s): Meteorites
Service(s): Meteorite Authentication

Energems Unlimited offers authenticated meteorite specimens for sale or trade; meteorite authentication, preparation, and restoration services; and they have designed and are currently producing an heirloom quality meteorite pendant featuring the Gibeon iron meteorite (that fell to earth in Namibia, Africa in 1836) available in sterling silver or 18k gold.

Energia Ltd.

631 South Washington Street
Alexandria, VA 22314
Phone: (703) 836-1999
Fax: (703) 836-1995
E-mail: energia@energialtd.com
Internet: www.energialtd.com

Product(s): Spacecraft
Service(s): Space Launch Support

Energia Ltd. is the American office for the Rocket Space Corporation (RSC) Energia, a Russian space corporation. Energia Ltd. was formed in February of 1992 specifically to assist RSC Energia in marketing its capabilities to potential international customers.

American office for the Rocket Space Corporation (RSC) Energia, a Russian space corporation

RSC Energia, Russia's oldest space organization, operates the Mir Space Station, and manufactures the Soyuz TM manned spacecraft and Progress M automated cargo spacecraft. RSC Energia is also the manufacturer of the Block DM upper stage for the Proton booster and new Sea Launch/Zenit-3 program. The company is the prime contractor for the Russian Space Agency segment of the International Space Station which is headed by NASA.

Equatorial Platforms

11065 Peaceful Valley Road
Nevada City, CA 95959
Phone: (916) 265-3183
E-mail: tomosy@nccn.net
Internet: www.Astronomy-Mall.com

Product(s): Equatorial Platforms

For Dobsonian telescope owners, Equatorial Platforms offers a selection of drive systems that will provide instant motorized tracking for a full hour at a time. Prices start at $850 (for the compact model designed for Dobsonians up to 12.5"). Write, call or e-mail for a full brochure.

Europtik, Ltd.

P.O. Box 319
Dunmore, PA 18512
Phone: (717) 347-6049
Fax: (717) 969-4330
E-mail: europtik@ptd.net
Internet: www.europtik.com

Product(s): Binoculars, Telescopes

Specializing in high end telescopes and custom observatory telescopes

Specializing in high end telescopes and custom observatory telescopes for amateur and institutional customers, Europtik's product line includes Intes, Intes Micro, Aries, Takahashi, Parallax, Pentax, Asko, Byers, OGS, Optomechanic Research, JSO, Lichtenknecker, AOK, UJI, and Telestar. Telescope brochures are free. A full catalog is available for $6 U.S./$8 foreign.

Exploration Software

P.O. Box 961
Groton, MA 01450-0961
Phone: (508) 649-4748
E-mail: hlynka@tiac.net
Internet: www.tiac.net/users/hlynka

Product(s): Computer Software

Exploration Software has produced a program called *On Top of the World* (for Windows) which allows you to fly around a three-dimensional model of the Earth as its position and sunlit side are correctly displayed for the time and date. Though the program is more geography-oriented, it does allow one to view the Earth from a space prospective, seeing how the sun lights the Earth as it rotates and how the cities look at night, among other features.

Galaxy Photography

Jason Ware
P.O. Box 835554
Richardson, TX 75083
E-mail: jasonw@galaxyphoto.com
Internet: www.galaxyphoto.com

Product(s): Photographs

Jason Ware's astrophotos are original color photographs, not poster reproductions. They are printed on high contrast Kodak or Fuji paper and are ready for framing, with sizes up to 16x20 inches. The current price for a 16x20 print is $20 plus shipping and handling. Jason's astrophotos have appeared in such prominent national magazines as *Sky & Telescope Magazine*, *Astronomy Magazine*, and the *Astronomy Observers Guide*. All products come with a 30-day money back guarantee.

Photos from top to bottom:

Comet Hyakutake
Flame Nebula
Andromeda Galaxy

Photo: Jason Ware

Photo: Jason Ware

Photo: Jason Ware

Historic Space Systems

12950 Tiger Valley Road
Danville, OH 43014
Phone: (614) 599-6779
E-mail: fongheis@genesys.net
Internet: www.space1.com

Photo: Historic Space Systems

Product(s): Spacecraft Exhibits and Simulations

Historic Space Systems makes space flight history come alive with spacecraft exhibits that launch imaginations! They offer realistic and historically accurate spacecraft exhibits that are carefully researched using 1,000's of actual spacecraft engineering drawings and other technical data. This is combined with CAD/CAM technology and skillful craftsmanship to create an exhibit that will make visitors feel as if they are seeing the actual spacecraft in flight! Exhibits and simulations currently available include Mercury, Gemini, and Apollo, plus the Lunar Module, Lunar Rover, Space Shuttle, and International Space Station. Call John Fongheiser, President, for more information, or visit their web site.

International Optics Ltd.

P.O. Box 6475
Nashua, NH 03063-6475
Phone: (603) 595-7978
Fax: (603) 595-7978
E-mail: 73554.3420@compuserve.com
Internet: Astronomy-Mall.com

Product(s): Telescopes

International Optics specializes in beginner and intermediate range telescopes that feature all metal constructions. Imported from Europe and Asia.

International Space Academy

P.O. Box 542327
Merritt Island, FL 32954
Phone: (407) 634-5151
E-mail: league@aol.com
Internet:
members.aol.com/league/isa.html

Product(s): Astronautics Training

The International Space Academy offers a complete in-home course in Astronautics. The course includes eight learning modules consisting of: *Life Support Systems*, *Space Mission Planning*, *Space Habitat Design*, *Space Colony Design*, *Habitat Operations*, *Space Mission Support and Logistics*, *Planetary Exploration and Excursions*, and *Emergency Planning*. The academy's faculty and staff have decades of experience in space exploration and have worked in the manned space programs such as Mercury, Gemini, Apollo and the Space Shuttle. A catalog and application are available free upon request.

International Star Registry

34523 Wilson Road
Ingleside, IL 60041
Phone: (708) 546-5533
Toll Free: 1-800-282-3333

Product(s): Gifts, Name-A-Star

Here's a unique gift idea that last a lifetime . . . and beyond. Have a star named after someone special, or claim your own place among the heavens by naming a star after yourself. You'll be joining the likes of Fred Astaire, Johnny Carson, Harrison Ford, Mick Jagger, Leonard Nimoy, Princess Diana, and many more. The International Star Registry gift package includes a 12x16 inch parchment certificate with the name of your choice, date of registration, and telescopic coordinates of the star. You'll also receive a chart of the constellations and a larger, more detailed chart with the star you name circled in red. The gift package costs $45 plus shipping and handling. Call for a brochure, current price, and application form.

Internet Telescope Exchange

7151 Market Street
Wilmington, NC 28405
Phone: (910) 686-9617
Fax: (910) 686-0042
E-mail: cescoWEB@aol.com
Internet:
www.burnettweb.com/ite/index.html

Product(s): Binoculars, Night Vision Equipment, Telescopes, Telescope Accessories

Internet Telescope Exchange (ITE) is owned and operated by amateur astronomers and specializes in high quality optics imported from Russia. ITE's product line includes optics and optical accessories by Intes, Intes Micro, LOMO PLC, LZOS, Newcon Optic, and ZMOS, along with US-made tripods by the Davis & Sanford Division of Tiffen. A complete catalog is available on the company's web site and amateur astronomers can buy, sell or trade used items for FREE on the exchange page of ITE's web site.

Jim's Mobile, Incorporated

810 Quail Street, Unit E
Lakewood, CO 80215
Phone: (303) 233-5353
Fax: (303) 233-5359
Orders: 1-800-247-0304

Product(s): Computer Software, Telescopes, Telescope Accessories

Jim Burr started Jim's Mobile Incorporated (JMI) in 1983 with the goal of providing products that solve problems. Rather than copying other companies, or competing with companies already building quality products, JMI looks for needs that are not being ad-

Photo: JMI

dressed and fills the gap. Starting with a single product, the MOTOFOCUS™ in 1983, JMI has grown its product line to include a selection of Next Generation Telescopes (NGT), New Technology Telescopes (NTT), Next Generation Computers (NGC), Next Generation Focusers (NGF), MOTOTRAK™ drive correctors and accessories, and much more.

Nice looking catalog!

Knollwood Books

P.O. Box 197
Oregon, WI 53575-0197
Phone: (608) 835-8861
Fax: (608) 835-8421
E-mail: books@tdsnet.com

Product(s): Books

Specializing in used and out of print books on astronomy, meteorology and space exploration, Knollwood Books also offers a free search service for those hard to find titles. Five book catalogs are published each year and are available free upon request.

Learning Technologies, Inc.

40 Cameron Avenue
Somerville, MA 02144
Phone: (617) 628-1459
Fax: (617) 628-8606

Product(s): Educational Materials, Planetariums (portable)

Manufacturer of the STARLAB Portable Planetarium, Learning Technologies, Inc., has helped bring the stars and planets to countless children (and adults) across the country. The inflatable STARLAB dome is easy to store and transport, making it ideal for school systems, and for museums where permanent space might be an issue. The standard STARLAB dome is 16 feet in diameter, 10.5 feet high, weighs only 45 pounds, seats 30 students, and can be set up in ten minutes. Aside from astronomy lessons, the dome can be used to teach subjects as varied as earth science, history, art, mythology, navigation, and geology.

Photo: Learning Technologies, Inc.

Learning Technologies also manufactures a variety of *Project Star* hands-on science materials developed by the Harvard Smithsonian Center for Astrophysics. These teaching aids include such items as a Sun Tracking Plastic Hemisphere Kit, Celestial Sphere Kit, 3-D Constellation Kit, Solar System Scale Model Kit, Refracting Telescope Kit, Project Star Teacher's Sampler, and more. Call for a Project Star catalog, or for a brochure on the STARLAB Portable Planetarium.

Loch Ness Productions

P.O. Box 1159
Groton, MA 01450-3159
Internet: www.lochness.com

Product(s): Planetarium Productions

Loch Ness Productions is devoted exclusively to the production of shows, slides and music for planetariums in U.S. and overseas.

Lumicon

2111 Research Drive
Suites 4-5
Livermore, CA 94550
Phone: (510) 447-9570
Fax: (510) 447-9589

Product(s): Astrophotography Equipment, Telescopes, Telescope Accessories

Lumicon sells Meade, Celestron, and Takahashi telescopes, plus a wide variety of telescope accessories, op-tics, and astrophotography equipment (including an assortment of gas-hyper-sensitized film and equipment for hy-persensitizing your own film). Lumicon also manufactures its own Finder Scopes, filters, focusers, adapters, piggyback mounts, and the Sky-Vec-tor family of computer-controlled digi-tal setting circle systems.

Mag 1 Instruments

16342 W. Coachlight Drive
New Berlin, WI 53151
Phone: (414) 785-0926
E-mail: PSmitka@aol.com

Product(s): Telescopes

The PortaBall telescope, by Mag 1 Instruments, offers a unique design that is highly portable and easy to as-semble, going from the car to viewing in a few minutes. The PortaBall 12.5-inch was the first introduced by Mag 1 and after proving itself as a powerful, user-friendly telescope, it was followed by Mag 1's new PortaBall 8-inch.

Peter Smitka, owner of Mag 1 In-struments, is personally involved with the manufacturing of each telescope, and as an avid observer, he's gone to great lengths to produce a high qual-ity, large aperture telescope that meets his viewing standards. Prices start at $1995 for the PortaBall 8-inch, and $2670 for the PortaBall 12.5-inch (manufacturer-direct). A selection of optional accessories is available.

MakeWood Products

P.O. Box 1716
Colfax, CA 95713
Phone: (916) 346-8963
Internet: members.aol.com/
donm353259/index.html

Product(s): Books

MakeWood Products publishes as-tronomy related books such as *An Observer's Guide to Comet Hale-Bopp*, *A Decade of Comets - A Study of the 33 Comets Discovered by Amateur Astrono-mers Between 1975 and 1984*, and *Messier Marathon Observer's Guide*. All three are priced at $12 each, plus $3 s&h.

Malco Precision Products

Phone: (516) 253-0720 (M-F, 9:30 am to 5:30 pm EST)
Orders: 1-800 895-1298

Product(s): Telescope Accessories

Malco Precision Products manufactures parts for truss-style Dobsonians. The system designed by Malco has three major components and is designed to allow quick assembly and disassembly, while providing ease of storage and transportation. Call for details.

Michael Blood Meteorites

6106 Kerch St.
San Diego, CA 92115-6628
Phone: (619) 286-4837
Fax: (619) 286-3134
E-mail: mblood@access1.net
Internet:
  www.meteorite.com/Michael_Blood

Product(s): Computer Software, Meteorites

Specializing in whole specimens and small samples of the rarer meteorites for individual collectors & institutions. Michael Blood also offers the *Space Rocks* CD, which contains a catalog of all meteorites, updated through 1996 (including 1000 photographs). Other available items include the MBC-10 microscope, insects in both amber & copal, tektites & miscellaneous rare items of interest. A continually updated catalog is accessible at Michael's web site. Those without Internet access may write, fax or phone for a recent printout of the catalog, which will be mailed to them free of charge.

The Meteorite Exchange

P.O. Box 7000-455
Redondo Beach, CA 90277
E-mail: info@meteorite.com
Internet: www.meteorite.com

Product(s): Meteorites

Offering more than just meteorites, The Meteorite Exchange is very active on the Internet, spreading the latest news and announcements as they apply to meteorites and meteorite related organizations. You'll get information on impact craters, auctions, dealer catalogs, dealers, books, and more. Check them out.

MMI Corporation

2950 Wyman Parkway
Mail:
P.O. Box 19907
Baltimore, MD 21211
Phone: (410) 366-1222
Fax: (410) 366-6311
E-mail: mmicorp@aol.com
Internet: members.aol.com/mmicorp

Product(s): Books, Computer Software, Educational Materials, Models, Planetariums, Slides, Videos

MMI Corporation's extensive catalogs are touted as the world's most complete catalogs of teaching materials for astronomy and geology. Products include 35mm slides, videos, CD-ROMs, laserdiscs, murals, software, celestial, earth, and planet globes, astronomical models of the sun, moon, and planets, spacecraft models, manuals, books, portable and permanent planetariums, and more.

Astronomy and geology catalogs are available free to U.S. educators, $5 (US$) per catalog for other domestic or Canadian addresses, $10 (US$) per catalog for all foreign addresses.

Moonbeam Meteorites

P.O.Box 9683
Ogden, UT 84409
Phone: (801) 627-1745
Fax: (801) 627-1745

Photo: David Leviton

Product(s): Meteorite Jewelry

Need a gift for someone special? How about something they probably don't have, like a ring, necklace, ear rings, belt buckle, pin or bracelet made from one of the rarest materials on earth . . . meteorite. The folks at Moonbeam Meteorites have been cutting and shaping meteorites since 1981 and take two expeditions each year to find more of the precious material. In addition to jewelry, they also manufacture knife blanks (for processing into blades) and can provide ½" x ½" chunks of Gibeon meteorite from Namibia, Africa for classrooms and collectors. All products are sold with a certificate of authenticity which gives a brief history of the meteorite, including its location and composition.

Mountain Skies Astronomical Society - Star Atlas

P.O.Box 1169
Lake Arrowhead, CA 92352
Phone: (909) 336-1699
E-mail: stargazersmail@worldnet.att.net
Internet:
 www.astro-msas.holowww.com

Product(s): Gifts

Mountain Skies Astronomical Society (MSAS) has a Star Atlas program where individuals can name a star and have it permanently recorded in the MSAS Star Atlas, which will be located at the MSAS Observatory and Science Center in Lake Arrowhead, California. Participants receive a certificate of record with the name and position of their star. Prices start at $30 and vary depending on the brightness of the star (there is a $3.50 shipping and handling charge for each certificate). As an extra touch, Star Atlas Commemorative Coins are also available in antique bronze or silver. This is a great gift idea for birthdays, Father's Day, memorials, births, Christmas, or just for the fun of it. See the Mountain Skies Astronomical Society listing under *Organizations* to learn more about the Observatory and Science Center you'll be supporting through your participation.

individuals can name a star and have it permanently recorded in the MSAS Star Atlas

New England Meteoritical Services

P.O. Box 440
Mendon, MA 01756
Phone: (508) 478-4020
Fax: (508) 478-5104
E-mail: nemsusa@delphi.com
Internet: www.meteorlab.com

Product(s): Books, Meteorites
Service(s): Meteorite Authentication, Appraisal, Preparation, Preservation

For those interested in collecting meteorites or purchasing an extremely unique gift, New England Meteoritical Services offers a wide variety of meteorites, comet and asteroid fragments, and even some extremely rare pieces of the planet Mars. If you're like most people and don't know a lot about meteorites, the New England Meteoritical Services catalog provides a very helpful introduction that includes meteorite terminology and a description of the primary types of meteorites (Chondrites, Iron, Stone, Mesosiderites, and others).

Many stunning meteorites are pictured in the catalog, such as a fragment of the Sikhote-Alin meteorite that impacted Eastern Siberia, Russia on February 12, 1947, fragments from the Odessa, Texas meteorite that impacted about 50,000 years ago, or fragments from the 30-meter diameter iron meteoriod that impacted Canyon Diablo "Meteor Crater" in Arizona.

New England Meteoritical Services also offers a variety of tektites, chondrules, fulgurite tubes (fused sand from lightning strikes), mantle material, diamond crystals, and trilobite fossils. Meteorite gift sets are available, as are presentation sets, collector's and classification sets, and a stunning meteorite sphere.

Photo: New England Meteoritical Services

A stunning piece of the Odessa meteorite that impacted near Odessa, Texas about 50,000 years ago.

North Star Systems

P.O. Box 99
Ellenburg Depot, NY 12935
Phone: (518) 594-7250

Product(s): Telescopes, Telescope Accessories

North Star Systems offers Dobsonian telescopes that are preassembled - no kits! All you have to do is lay the telescope in the mount, insert an eyepiece, and start observing. All North Star telescopes use a ventilated mirror mount, with the back and sides open to air circulation, allowing the mirror to attain temperature stability quickly and reliably. Available systems include a 6" f/8, 10" f/5.6, 8" f/6, 4.5" f/10 Baby Dob, and a variety of Newtonian reflector tube systems and Dobsonian mounts.

Observa-Dome Laboratories

371 Commerce Park Drive
Jackson, MS 39213
Phone: (601) 982-3333
Fax: (601) 982-3335
E-mail: odl@misnet.com/odl
Internet: www.misnet.com/odl

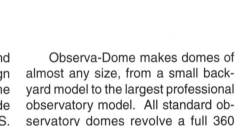

Laboratories, Inc.

Product(s): Domes

Over forty years of experience and expertise have gone into the design of Observa-Dome products. The company has customers worldwide and in most states, including the U.S. Naval Academy, Kitt Peak National Observatory, Smithsonian Astrophysical Observatory, Kennedy Space Center, European Southern Observatory, the Adler Planetarium (Doane Observatory), Cape Canaveral, and more.

Observa-Domes are primarily constructed of aluminum because it has a mass one third that of steel or fiberglass and because it does not have to be coated to eliminate rust. Though steel and fiberglass are less expensive than aluminum, maintenance costs make them less practical. The company states that they do make Observa-Domes of steel, stainless steel, fiberglass, and copper if that's what the customer wants.

Observa-Dome makes domes of almost any size, from a small backyard model to the largest professional observatory model. All standard observatory domes revolve a full 360 degrees and are equipped with a biparting shutter system. Contact Observa-Dome for information and a brochure.

Photo: Observa-Dome

Observatory Techniques

Mike Otis, Editor
1710 SE 16th Avenue
Aberdeen, SD 57401-7836
Phone: (605) 226-1078
E-mail: otm@midco.net
Internet: ourworld.compuserve.com/
homepages/observatory

Product(s): Magazine

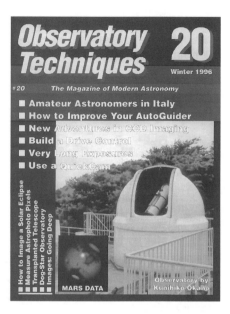

Observatory Techniques 20
Winter 1996

#20   *The Magazine of Modern Astronomy*
■ Amateur Astronomers in Italy
■ How to Improve Your AutoGuider
■ New Adventures in CCD Imaging
■ Build a Drive Control
■ Very Long Exposures
■ Use a QuickCam

How to Image a Solar Eclipse
Measure Astrophoto Pixels
Transplanted Telescope
Dog-Star Observatory
Images: Going Deep
MARS DATA   Observatory by Kunihiko Okano

Observatory Techniques is a quarterly magazine that covers a wide range of subjects of interest to amateur astronomers. A recent edition included articles such as *How to Study Drift Scan Images*, *How to Improve Your Autoguider*, *How to Image a Solar Eclipse*, *Amateur Astronomers in: Italy*, and *Adventures in Imaging*. This is truly a magazine by amateur astronomers, for amateur astronomers. Annual subscription: $28 ($36 Canada/Mexico, $38 Overseas - surface).

Orion Telescopes & Binoculars

P.O. Box 1815
Santa Cruz, CA 95061
Offices: (408) 763-7000
Catalog Requests: 1-800-447-1001
Fax: (408) 763-7017
E-mail: sales@oriontel.com
Internet: www.oriontel.com

Photo: Orion Telescopes & Binoculars

Product(s): Binoculars, Books, Computer Software, Spotting Scopes, Telescopes, Telescope Accessories

Orion Telescopes & Binoculars is the manufacturer and dealer of a wide variety of optical instruments and accessovries for amateur astronomers and outdoor enthusiasts. Brands include Orion, Celestron, Tele Vue, Bushnell, Fujinon, Swarovski, Kowa, Nikon, Pentax, Bogen, Zeiss, Leica, Canon, and others. The company also offers telescope-making components and an expert optical repair service. In business since 1975, Orion has a free 100-page, full-color catalog which is available upon request. An online catalog is also available.

Palo/Haklar Multimedia

3740 Overland Avenue #H
Los Angeles, CA 90034
Phone: (310) 558-8839
Fax: (310) 558-8836
E-mail: pha@earthlink.net

Product(s): Computer Software

Palo/Haklar currently produces and distributes two space-related software titles. *Voyage Through the Solar System 2.0* contains detailed information from NASA space probes, full motion video clips from the Jet Propulsion Laboratory, animations, photos, and slide shows with music and full narration. Palo/Haklar's second space title, *The Last Frontier*, allows the user to relive the most significant moments in the space race, including a behind-the-scenes look at the astronauts and the space program. The CD includes over an hour of video footage, details of each mission, and more.

Both *Voyage Through The Solar System* and *The Last Frontier* are available in Windows and MacIntosh versions and are suitable for ages 8 to adult. Lab packs are available. Palo/Haklar has an agreement to provide up to 150,000 copies of *Voyage Through The Solar System* for packaging with college astronomy text books - a significant endorsement for this software title. *The Last Frontier* sells for $34.95 and *Voyage Through The Solar System* sells for $45.95.

Parallax Multimedia, Inc.

1233 Saddle Court
Auburn, CA 95603
Phone: (916) 887-0225
Fax: (916) 887-8729
E-mail: parallax@quiknet.com
Internet: www.quiknet.com/~parallax

Product(s): Computer Software

GalaxyGuide, A Journey through the Constellations is available from Parallax Multimedia. This CD covers 38 constellations in both the northern and southern hemispheres. It includes accurate star maps, a "virtual telescope" where users can click on the star map and zoom to telescopic photographs from the world's major observatories, a section that explores the mythology of each constellation, pronunciation of constellation and star names, a complete astronomy dictionary, two games, and more. The pro-

gram is easy to set up and use and is suitable for adults and children age 13 and older. *GalaxyGuide* sells for $34.95, plus $5 shipping and handling (CA residents add 7.25% sales tax).

Parks Optical

P.O. Box 716
Simi Valley, CA 93065
Phone: (805) 522-6722
Fax: (805) 522-0033

Product(s): Binoculars, Telescopes, Telescope Accessories

Parks Optical manufactures a wide variety of quality telescopes, binoculars, spotting scopes, mounts, mirrors, filters, and other accessories. Telescopes vary in design and capability, covering virtually all observing requirements and budgets. Parks manufactures the Observatory Series, Superior Series, Precision Series, Astrolight Series, Nitelight Series, and Newtonian-Cassegrain Series. Binoculars range from the Parks Deluxe Giant Binoculars down to the 8x25 compact unit and the opera glasses.

Photo: Parks Optical

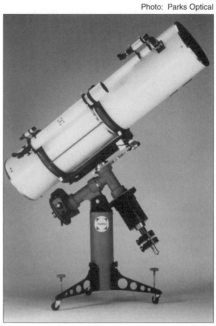

16" f/5 Observatory Series Telescope

Photo: Parks Optical

12.5" f/5 Superior Series Telescope

Parks manufactures the Observatory Series, Superior Series, Precision Series, Astrolight Series, Nitelight Series, and Newtonian-Cassegrain Series.

Pegasus Optics

RR 5, Box 502
Huntsville, AR 72740
Phone: (501) 738-1650
Fax: (501) 738-1650
E-mail: jhall@cswnet.com
Internet: www.icstars.com/pegasus

Product(s): Telescope Mirrors

Pegasus Optics offers Newtonian telescope mirrors that are 2" thick, precision-annealed, and fine-ground on the back surface (thickness and diametral tolerance is .125"). Enhanced coatings are available. Mirror precision for F/4.2 or longer f-ratio is diffraction-limited, F/4 and below as defined by Danjon-Couder conditions for wavefront error and diffraction disk diameter. Surface errors average less than 1/20 wave p-v, wavefront average less than 1/10 wave p-v (all diameters/F-ratios), and relative transverse aberration.

Each mirror is etched with purchaser's name, serial number, and optician signature. Traceability, quality data and certification package for the serialized mirror are provided to customers and copies are maintained in an historical file. Terms are half down with the balance due prior to delivery. Catalog: $1 U.S./$2 foreign.

Personal MicroCosms

8547 East Arapahoe Road
Suite J-147
Greenwood Village, CO 80112
Phone: (303) 753-3268
E-mail: EricTerrell@juno.com
Internet: users.aol.com/ericb98398/index.html

Product(s): Computer Software

Offering the astronomical shareware program *Astronomy Lab* for Windows 3.1, Windows NT 3.5, and Windows 95. *Astronomy Lab* produces 7 movies that simulate a host of astronomical phenomena, 15 graphs that illustrate many fundamental concepts of astronomy, and 14 printed reports that contain predictions of the most important astronomical events. All movies, graphs, and reports are customized for the user's time zone and location. *Astronomy Lab* can be ordered by sending a 3.5" or 5.25" high density floppy disk and a self-addressed, postpaid envelope to Personal MicroCosms (specify which version of Windows you want). To get a *registered* copy, which includes access to Personal Microcosms' customer support telephone number and notification of future versions, send $15 plus $2.50 postage in the U.S. and Canada ($10 overseas).

Pocono Mountain Optics

104 N.P. 502 Plaza
Moscow, PA 18444
Phone: (717) 842-1500
Fax: (717) 842-8364
Orders: 1-800-569-4323
E-mail: pocmtnop@ptdprolog.net
Internet: astronomy-mall.com

Product(s): Binoculars, Spotting Scopes, Telescopes, Telescope Accessories

Dealer and distributor for all major brands of telescopes, binoculars, and accessories, Pocono Mountain Optics handles brands such as Meade, Tele Vue, Celestron, Bausch and Lomb, Kowa, Fujinon, Leica, Zeiss, Takahashi, Pentax, Losmandy, JMI, Edmund Scientific, Nikon, Orion, and others. Showroom is stocked with over 45 scopes on display.

Pulsar Publishing

56 Kimbolton Close
Lee, London SE12 OJJ
United Kingdom
Phone: 0181-488-1701
E-mail: jmwebb@dircon.co.uk
Internet: www.users.dircon.co.uk/
~jmwebb/pulsar.htm

Product(s): Computer Software

The latest offering from Pulsar Publishing is *The Windows Dictionary of Astronomy*, second edition. This Windows compatible program is a comprehensive astronomical dictionary which incorporates over 2000 terms with associated terms and uses hypertext links to move to related educational articles. A table viewer is provided which includes tables on Constellations, Planetary Conjunctions, Satellites, Orbital and Physical Characteristics, Meteor Showers, White Dwarfs, and more.

The Windows Dictionary of Astronomy, *now in its second edition*

R.V.R. Optical

P.O. Box 62
Eastchester, NY 10709
Phone: (914) 337-4085
Fax: (914) 3377-4085

Product(s): Binoculars, Flat Field Cameras, Telescopes, Telescope Accessories

Offering an impressive line-up of telescopes, mountings, and other astronomical instruments, R.V.R. Optical can provide Newtonians up to 20-inch, Cassegrains up to 138-inch, Newtonian/Cassegrains up to 60-inch, Schmidt Cassegrains up to 15 3/4-inch, giant binoculars, specialized solar telescopes, night vision scopes, antique telescopes, and much more. If you're looking for a high-end telescope,

Photo: R.V.R. Optical

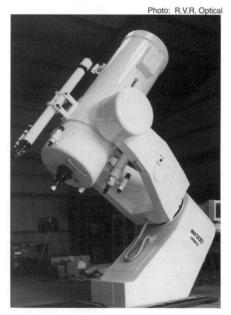

ASKO SG500 Observatory Telescope

how about a complete 24" Cassegrain telescope system for $147,600 (allow one year for manufacture). A 30" model is available for $248,900 and a 40" model can be purchased for $334,000 (prices subject to change). Of course, R.V.R. Optical also offers smaller and less expensive models.

Rainman Software

100 Shale Place
Charlottesville, VA 22902-6402
Orders: 1-888-261-1686
Tech Support: (804) 984-2808
Fax: (804) 984-1241
E-mail: ngcview@rainman-soft.com
Internet: www.rainman-soft.com
Internet FTP: ftp.rainman-soft.com

Product(s): Computer Software

Rainman Software offers *NGCView* for Windows (available in both 16-bit and 32-bit versions), which combines a powerful astronomical database engine and a graphical interface to create an easy to use tool for planning and recording observations. The program includes many useful filtering and sorting algorithms not found in other products, and provides a wide variety of graphs, charts, coordinate system and numerical formatting options. An observing log database allows you to manage systematic surveys, photographic observations and CCD observations. The software can also create finder charts for all objects in one step. Cost is $49.95 (new user), or $25 upgrade with free interversion online updates.

Rivers Camera Shop

454 Central Avenue
Dover, NH 03820
or
69 North Main Street
Rochester, NH 03867
Phone: (603) 742-4888
Orders: 1-800-245-7963
Fax: (603) 742-5456
E-mail: Rivers@ttlc.net

Product(s): Binoculars, CCDs, Telescopes, Telescope Accessories

In business since 1930, Rivers Camera Shop has two stores in New Hampshire that do business across the country, and the largest in-store telescope display on the East Coast. Their catalog, *Rivers Astronomy Guide*, offers a nice assortment of telescopes, including Meade, Celestron, Tele Vue, and others.

Rivers' Specialized Astronomy Catalog

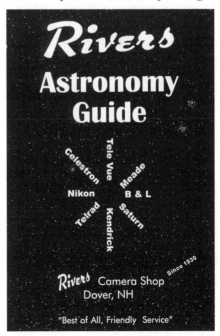

Jay Ryan
Astronomical Cartoonist

P.O. Box 609118
Cleveland, OH 44109
E-mail: starman@cyberdrive.net
Internet: www.skywise.com

Product(s): Astronomical Cartooning

Jay Ryan is the creator of *Starman*, a monthly cartoon feature which appears in the newsletters of over 150 astronomy clubs, museums and park districts around the United States and the world. Ryan's work is guided by the premise that astronomy is an inherently "visual" subject. The appearances of the sky are comprehended only through the eyes. However, astronomy is typically taught using largely non-visual media, such as text accompanied by sparse smattering of pictures. The cartoonist views the "comic strip" as an excellent medium for exploiting the visual advantages of illustration in order to teach the visual elements of astronomy.

In addition to *Starman*, Jay Ryan is the creator of *Cycles: An Illustrated Introduction to Astronomy and Time*. *Cycles* is a 32-page cartoon booklet that explains the three fundamental measures of time, The Day, The Month, and The Year, and how these measures are derived from the ancient Cycles of the Sun and Moon. *Cycles* is published by the Astronomical Society of the Pacific (ASP). The cartoonist is also selling copies of *Cycles* through his P.O. Box for $2.50, plus $1.00 postage and handling ($3.50 total, US dollars).

Jay Ryan is also the creator of *SkyWise*, an astronomy comic strip which appears each month in *Sky & Telescope* magazine. Unlike *Starman*, *SkyWise* is larger and in full color! Ryan's work has also appeared the ASP teacher's newsletter, *Universe in the Classroom*, and also in the ASP *Mercury*. Ryan hopes to someday produce a series of book-length astronomy comic books that will teach a variety of topics in practical visual astronomy.

Science & Art Products

24861 Rotunde Mesa
Malibu, CA 90265
Phone: (310) 456-2496
Fax: (310) 456-0728
E-mail: fletcher@scienceandart.com
Internet: www.scienceandart.com

Product(s): Gift Items, Souvenirs

Specializing in gift and souvenir items that are manufactured with planetarium, observatory, and museum gift shops in mind, Science & Art Products carries a wide selection of astronomy-related products, including posters, slide sets, greeting cards, t-shirts, magnets, calendars, photo bookmarks, children's books, and more. Products are available either retail or wholesale.

Scope City

730 Easy Street
Simi Valley, CA 93065
Phone: (805) 522-6646
Fax: (805) 582-0292

TELESCOPES • BINOCULARS • SPORT OPTICS

Product(s): Binoculars, Books, Telescopes, Telescope Accessories

Scope City sells telescopes, spotting scopes, binoculars, microscopes, star charts, books, and related accessories. All major brands are carried, including Celestron, Parks, Meade, Tele Vue, Questar, Zeiss, Leica, Steiner, Dr. Optics, Bausch & Lomb, Minolta, Pentax, Nikon, and others. Scope City has retail locations in Simi Valley, Sherman Oaks, Costa Mesa, and San Diego, California. They also ship products nationally and internationally.

Seattle Support Group

20420 84th Avenue South
Kent, WA 98032
Phone: (253) 480-1001
Fax: (253) 480-1006
E-mail: sales@ssgrp.com

Product(s): Computer Software, Photographs

Photo Courtesy of Seattle Support Group

For astronomy clubs and businesses looking for a way to spice up their newsletter or brochure, Seattle Support Group publishes two photo CDs entitled Space I and Space II. These CDs contain royalty-free, printable photos of space shuttles, planets, satellites, blastoffs and landings, astronauts, views of earth, and the Hubble Space Telescope. All images were scanned in 24-bit color at 300 dpi resolution and color-corrected. Each CD contains 200 images, offering enough variety to keep your correspondence looking great for years to come.

Sky Publishing Corporation

P.O. Box 9111
Belmont, MA 02178-9111
Phone: (617) 864-7360
Orders: 1-800-253-0245
Fax: (617) 864-6117
Editorial Fax: (617) 576-0336
E-mail: skytel@skypub.com
Internet: www.skypub.com

Product(s): Magazines

Publisher of the prominent *Sky & Telescope* magazine, the new annual publication *SkyWatch*, and the *Sky Publishing Catalog*, Sky Publishing Corporation has been a leading source of accurate and up-to-date information about astronomy and space science since 1941.

Sky & Telescope, Sky Publishing's flagship publication, appeals to the full spectrum of astronomy enthusiasts, from the novice ready to purchase a first telescope, to the enthusiastic amateur looking to enhance observing skills and equipment, to the professional desiring to keep up with this dynamically changing field. The magazine is filled with articles written to satisfy both technically savvy readers and those who require clear, descriptive language and superb graphics. In addition to the rich imagery throughout the magazine, the Gallery section showcases stunning works by amateurs using conventional and electronic cameras. After browsing through a copy, you'll see why *Sky & Telescope* is the magazine of choice for a host of amateur astronomers around the world and why more than 190,000 people read *Sky & Telescope* each month.

A one year (12 issue) subscription to *Sky & Telescope* costs $36 ($46 in Canada, $50 for all other international subscriptions). Participating astronomy clubs can receive the magazine for $27, while also enjoying a 10% discount on all purchases of astronomy-related books and products from the Sky Publishing Catalog. Students can also get *Sky & Telescope* at this reduced price if they provide a letter on school stationery attesting to student status and signed by a school official.

Sky Publishing recently introduced *SkyWatch '97*, a new annual publication written for anyone with a general interest in astronomy and space flight. Geared mainly toward beginners, it will also appeal to amateur astronomers and other outdoor enthusiasts who want to enhance their enjoyment of the night sky and make informed decisions about buying binoculars, telescopes, astronomy software, and related products. Ask for *SkyWatch* at your local newsstand or order directly from Sky Publishing (and save 20%).

Finally, Sky Publishing offers an excellent color catalog featuring over

Leo's Finest Galaxies and Double Stars • Sunspots That Changed the World

SKY & TELESCOPE

April 1997 $3.95
$5.25 CAN.

Is Radio Astronomy Doomed?

Top Binoculars for Stargazing

Comet Hale-Bopp

Where and How To See It

Shigemi Numazawa's Vision of Hale-Bopp

200 books, star atlases, observing guides, software, maps, globes, videos, posters, planispheres, slides, and CD-ROMs specially selected by the editors of *Sky & Telescope*. If you would like a printed copy of this catalog or would like one sent to a friend, send your request by e-mail to catalog@skypub.com, or call the toll-free or regular phone number.

Skywatcher's Inn

Benson, AZ
c/o 420 S. Essex Lane
Tucson, AZ 85711
Phone: (520) 745-2390
Fax: (520) 745-2390
E-mail: vegasky@azstarnet.com
Internet:
www.communiverse.com/skywatcher

Service(s): Accommodations

Located 47 miles east of Tucson, and not far from the historic town of Tombstone, Skywatcher's Inn bills itself as "The Arizona Astronomy and Nature Retreat Bed and Breakfast." and caters mostly to amateur astronomers, birdwatchers, and nature lovers.

The main attraction at Skywatcher's Inn is the Vega-Bray Observatory (see separate listing in the *Places of Interest* section), a privately owned amateur astronomy observatory dedicated to public education. The site, co-located with the inn, is situated at 3800 feet elevation and houses telescopes ranging in size from 6 to 20 inches, CCD imaging equipment, a small science museum and small planetarium. Classes in astronomy for beginners and advanced amateurs are offered by appointment.

Photo: Skywatcher's Inn

Skywatcher's Inn features the Vega-Bray Observatory

Software Bisque

912 Twelfth Street, Suite A
Golden, CO 80401-1114
Phone: (303) 278-4478
Orders: 1-800-843-7599
Fax: (303) 278-0045
Internet: www.bisque.com/thesky

Product(s): Computer Software

Software Bisque has developed some amazing software called *TheSky* that claims to put "the power of the universe" in your hands. As you start to work with *TheSky* you quickly discover this is no idle claim. This pow-

amazing software

Screen capture from TheSky

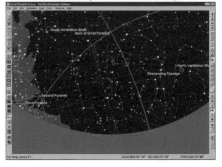

Screen Capture from TheSky

erful program actually serves as a personal planetarium, allowing even first-time astronomers to travel through the galaxies with ease. First released in 1984, *TheSky* is in its fourth version, now taking advantage of 32-bit technology and Windows 95 (the program also works with Windows NT and Windows 3.1x, and a DOS version is available).

Another outstanding program by Software Bisque is *SkyPro* CCD astronomy and image processing software. This program orchestrates a CCD camera, telescope and computer to download celestial images into a computer's memory. Used in conjunction with *TheSky*, your CCD astronomy will be easier than ever. While *SkyPro* is currently a 16-bit application, Software Bisque is releasing a 32-bit application called *CCDSoft* that will include many upgraded features from *SkyPro*.

Aside from these programs, Software Bisque wrote the control software for *RealSky* astronomy software, which they also sell. It should be noted that *TheSky* is the only astronomy software which supports *RealSky* software. This is an interesting, involved company with some great products. Check out their web site!

Sovietski Collection

P.O. Box 81347
San Diego, CA 92138
Phone: 1-800-442-0002
International: (619) 294-2000
Fax: (619) 294-2500
E-mail: fulcrum@sovietski.com
Internet: www.sovietski.com

Product(s): Binoculars, Optics, Telescopes

Sovietski Collection has a free 32-page catalog featuring such items as their giant (15x110) observation binoculars, 4.5" and 6" Newtonian reflector telescopes with clock drives from a premier Russian military optics factory, aviation/space collectibles and more. Four pages in the company's new catalog will be devoted to aviation and space collectibles.

Space Age Collectibles

14494 Neptune Road
Seminole, FL 33776
Phone: (813) 595-0788
Fax: (813) 596-1345
Orders: 1-800-800-2001
E-mail: spaceage@gate.net
Internet:
 www.spaceage-collectibles.com

Product(s): Clothing, Collectibles, Gift Items, Models, Patches, Videos

Space Age Collectibles provides an opportunity to own your share of outer space exploration from their impressive collections which cover the NASA programs for Mercury, Gemini, Apollo, Skylab, Space Shuttle, Launch Boosters, Earth Satellites, Planetary Explorers, and Hi-Tech aircraft. Collections are also available on international programs for the Space Station, Earth Satellites, and Planetary Explorers. Space Age Collectibles supports individuals, groups, museums, teachers and students with high quality space items and personalized service whenever possible. Custom orders are welcome.

The company's world wide direct marketing and Internet catalog provide a way for you to start or add to your collection now. Space Age Collectibles catalog includes sections on Mission Patches and Decals; Space Caps, Shirts and other Apparel; Mission Pins, Jewelry and Coins; Space Prints, Videos and CDs; Space Philatelic Covers and Services; Space Models and Cups; and Space Educational and Funtastic Gifts. They make it easy to ask questions and can process your order (or question) by e-mail, fax, mail, or toll-free calling. Major credit cards accepted.

Space Camp

U.S. Space & Rocket Center
One Tranquility Base
Mail: P.O. Box 070015
Huntsville, AL 35805-3399
Phone: (205) 837-3400
Space Camp: 1-800-63 SPACE
Fax: (205) 837-6137
Internet: www.spacecamp.com

more than 230,000 trainees have attended U.S. Space Camp

Service(s): Education

Since opening in 1982, more than 230,000 trainees have attended U.S. Space Camp from all 50 states and around the world. In addition to the U.S. Space Camp in Huntsville, Alabama, Space Camp programs are now available in Mountain View, California and Titusville, Florida. The training program is the same at all locations, with most youth camps lasting five days, beginning on Sunday and ending on Friday. A Parent/Child program is available which begins on Friday and ends on Sunday.

Over the years, Space Camp has added training programs to accommodate older students. Programs now include: Space Camp (for trainees in grades 4 through 6), Space Academy (grades 6 through 8), Advanced Space Academy (grades 9 through 12), Adult Space Academy, Space Academy for Educators, Parent/Child Space Camp, and Corporate Space Camp. Participants don't just learn what astronauts do, they actually do what real astronauts do. Like strapping into the 5 Degrees of Freedom simulator to train for a space-walk, or experiencing what it's like to move around on the moon in the 1/6th Gravity Chair.

Sessions are designed around simulated space missions conducted in Space Shuttle orbiter mockups. Trainees learn the basics of shuttle operation, the science and history of the space program, leadership skills and teamwork. Each trainee is assigned a role during the simulated shuttle mission. Some "fly" the orbiter, while others do experiments or "space walk" activities. The rest of the team members handle responsibilities in mission control.

Trainees in Huntsville, Alabama also tour NASA's Marshall Space Flight Center, those at Space Camp in Florida tour NASA's Kennedy Space Center, and trainees at Space Camp California tour NASA's Ames Visitor Center at Mountain View.

Tuition for the week-long Space Camp and Space Academy is $675. Advanced Space Academy tuition is $875, the Parent/Child weekend is $300, and the weekend Adult Space Camp is $500. Tuition is subject to change so call for current prices and program details.

Space Country Souvenirs

1126 West Ocean Avenue
Lompoc, CA 93436
Phone: (805) 735-1322
Fax: (805) 737-4472
E-mail: spaceco10185@www1.utech.net
Internet: www.dmatrix.com/spaceco

Product(s): Collectibles, Novelties, Souvenirs

A leading producer and distributor of space program patches, caps, emblems, decals, buttons, T-shirts, and other memorabilia, Space Country Souvenirs specializes in custom emblems and emblematic jewelry. They also have a large selection of specialty patches of Soviet and Russian, Military, commercial, and satellite launch programs.

Space Craft International

P.O. Box 61027
Pasadena, CA 91116-7027
Phone: (626) 793-4300
Fax: (626) 793-4233
E-mail: dd@scikits.com
Internet: www.scikits.com

Product(s): Spacecraft Models

Here's a neat product. Space Craft International (SCI) manufactures lightweight, laser-cut paper model kits of various space vehicles, such as the Mars Global Surveyor. These Space Craft™ Science Kits are designed so that you learn about the vehicle as you assemble it. For example, the Mars Global Surveyor has internal parts and markings which cannot be seen when assembly is complete, but by including these parts, you learn the vehicle inside and out as you build it. Other kits include the spacecrafts

Assembled Model of the Mars Global Surveyor

Galileo, Voyager, Magellan, the Hubble Space Telescope, and the Keck Telescope. The cost is $14 per kit, with reduced rates for orders of 4 or more. Shipping in the U.S., Canada, and Mexico is free, but California residents must add 8% sales tax.

Space Images

P.O. Box 701567
San Antonio, TX 78270-1567
Phone: 1-800-877-8915
Fax/Phone: (210) 499-4504
E-mail: simages@netxpress.com
Internet: www.SpaceImages.com

Photo: Jeff Hester and Paul Scowen (Arizona State University), and NASA

Product(s): Photographs

Space Images produces quality hard copy photographs from various NASA programs, particularly the Hubble Space Telescope, the Space Shuttle, and the Apollo missions. Photos are reasonably priced and available in four standard sizes.

Gaseous Pillars in the Eagle Nebula, taken from the Hubble Space Telescope

Space Works, Inc.

103 N. Whiteside
Hutchinson, KS 67501
Phone: (316) 662-9301
Fax: (316) 665-4043

Space Works is currently restoring "Odyssey," the Apollo 13 Command Module.

Product(s): Space Hardware Reproductions
Service(s): Space Hardware Restoration

Space Works, Inc., a wholly-owned subsidiary of the Kansas Cosmosphere and Space Center (KCSC), operates the world's only permanent space artifact restoration and replication facility. During the past 14 years, more than 50 major restoration and replication projects have been conducted by the firm's staff. Originally developed to provide restoration and exhibit support to the KCSC's vast space artifact collection, Space Works' unique expertise is now heavily demanded by groups throughout the world. As its services have grown, so has the company's expertise. Today, Space Works claims the distinction of being the world's recognized leader in the fields of research, restoration and preservation of American space artifacts, as well as the world's leading producer and supplier of museum-quality replicas of American and Russian space hardware.

Stargazer Steve

1752 Rutherglen Cres.
Sudbury, Ontario
P3A 2K3 Canada
Phone: (705) 566-1314
Internet: ww2.isys.ca/stargazer

Product(s): Telescopes

Steve Dodson is an award-winning telescope maker who has designed an affordable reflecting telescope with new stargazers in mind. The Sgr-3 is a 3-inch reflecting telescope that comes with light but sturdy wooden fork altazimuth mounting. The scope is capable of 150x magnification and sells for a very affordable $229 USD, plus $29 shipping and handling (subject to change).

Photo: Stargazer Steve

Stargazer Steve's inexpensive 4 1/4-inch Deluxe Reflector Kit

Star Hill Inn

Sapello, NM 87745
Phone: (505) 425-5605
Internet: www.starhillinn.com

Service(s): Accommodations, Telescope rentals

Open year round, Star Hill Inn sits on 195 private acres and offers seven cottages with fully equipped kitchens, private porches, and cozy fireplaces. Catering to astronomers and bird-watchers, the inn is located at one of the best dark-sky locations in the United States. Rental telescopes in a wide range of sizes and types are available on a nightly or weekly basis and private observing assistance is available at an hourly rate (advance reservations requested). Powered piers and tripod space are available for guests who bring their own equipment. Minimum stay at Star Hill Inn is two nights. Reservations are required.

Stellar Products

7387 Celata Lane
San Diego, CA 92129
Phone: (619) 538-8362
E-mail: info@stellarproducts.com
Internet: www.StellarProducts.com

Product(s): Photographic and Imaging Accessories, Adaptive Optics

Stellar Products manufactures the AO-2, the only adaptive optics system for amateur astronomers. This system reduces atmospheric turbulence, telescope jitter, and clock drive errors, resulting in near diffraction-limited images of planets. See their web site for more information.

Sunset Cliffs Merchandising Corporation

P.O.Box 7518
San Diego, CA 92167
Phone: (619) 692-8196
Fax: (619) 692-8199

Product(s): Binoculars, Optics

Sunset Cliffs is the U.S. distributor for Optolyth-Optik, a premier German optical house specializing in scopes and binoculars. The company recently introduced an all new 15x63 BGA roof prism binocular that features full rubber armor, nitrogen filling, long eye relief, phase corrected roof prisms, and Optolyth's famous "Ceralin-plus" multicoating on all optical surfaces.

Swayze Optical

700 NE 101st Avenue, #31
Portland, OR 97220
Phone: (503) 252-5009
Fax: (503) 335-3320
E-mail: swayze@europa.com

Product(s): Telescope Mirrors

Specializing in high quality telescope mirrors for amateur astronomers and telescope makers, Swayze Optical offers sizes from 12.5-inch to 30-inch with your choice of thickness, focal length, and coatings. Taking advantage of nearly twenty years of mirror-making experience, Steve Swayze puts his mirrors through a series of tests, including the foucault test, the Ronchi test, and the Ross Null test to insure the best possible surface smoothness and accurate correction. Call Steve or Bruce for prices and information.

Specializing in high quality telescope mirrors for amateur astronomers and telescope makers

Swift Instruments, Inc.

952 Dorchester Avenue
Boston, MA 02125
Mail:
P.O. Box 562
San Jose, CA 95106
Phone: (617) 436-2960
Fax: (617) 436-3232
Internet: www.swift-optics.com

Photo: Swift Instruments

Product(s): Binoculars, Spotting Scopes, Telescopes

Swift's Model 868 Achromatic Refracting Telescope

Swift Instruments has been providing quality optics since 1926. They import and manufacture items ranging from binoculars, rifle scopes, and telescopes to readers, magnifiers, and weather and marine instruments. An example of their work can be seen in their Model 868 Achromatic Refracting Telescope, which features a solid brass tube, equatorial mount, brass finder scope, and wooden tripod with triangle tray.

TECH2000

3349 SR 99 South
Monroeville, OH 44847
Phone: (419) 465-2997

Product(s): Telescope Accessories

TECH2000 allows you to take full advantage of your Dobsonian by giving it the "dead-on" tracking of an equatorial

Offering products like the Dob Driver II (a drive system for Dobsonians that has sold over 1000 units), TECH2000 allows you to take full advantage of your Dobsonian by giving it the "dead-on" tracking of an equatorial. The Dob Driver II costs $549 and comes with full installation and operation manuals, plus TECH2000's technical support. The system installs on scopes from 3" to 30" in a few hours with simple tools.

The Dob Driver Kit

Photo: TECH2000

Technical Innovations, Inc.

22500 Old Hundred Road
Barnesville, MD 20838
Phone: (301) 972-8040
E-mail: domepage@erols.com
Fax: (301) 349-2441

Product(s): Domes

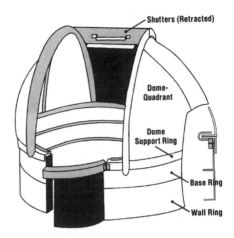

PD-10 (Not to Scale)

While making repairs to their 25-year old home-made dome in 1991, John and Meg Menke decided it was time to put their knowledge of observatory construction to work. The result was Technical Innovations, manufacturer of HOME-DOME and PRO-DOME observatories. Manufactured of fiberglass, their domes are designed for strength, appearance, and long life. Five models are currently available, from the six-foot HD-6S to the fifteen-foot PD-15. Electric shutter and rotation is included with the PD-15, and optional in all other models. Dome automation systems are available, including DOME-WIZARD, which uses a computer to control your telescope, CCD, and dome, and DOME-WORKS, a less sophisticated non-computer system. Technical Innovations now has domes installed at homes, schools, colleges, museums, and research sites across the United States and around the world.

Tele Vue

100 Route 59
Suffern, NY 10901
Phone: (914) 357-9522
Fax: (914) 357-9523
Internet: www.televue.com

Photo: Tele Vue

Nagler Type-2 Eyepieces

Product(s): Spotting Scopes, Telescopes, Telescope Accessories

Using skills acquired over almost two decades as an optical systems designer in support of NASA (specifically, the displays for the Gemini and Apollo Lunar Module spacecraft), Al Nagler founded Tele Vue Optics in 1977. From the beginning, his accomplishments have been impressive, to include multiple patents, such as his patented Nagler eyepiece and patented 4-element Multi-Purpose Telescope (MPT). He also manufactures the award-winning Renaissance Telescope, as seen on the front cover mounted on Tele Vue's sturdy Gibraltar mount.

Tele Vue's line up includes telescopes, a wide variety of eyepieces (including Panoptic™ wide-angle eyepieces), mounts, sights, spotting scopes, and Sky Tour, which adds computer technology to the Gibraltar Mount to help locate 2000+ interesting and observable items. Send $3 for a complete literature package, including reviews and articles.

Telrad, Inc.

P.O. Box 6780
Pine Mountain Club, CA 93222
Phone: (805) 242-5421

Product(s): Telescope Accessories

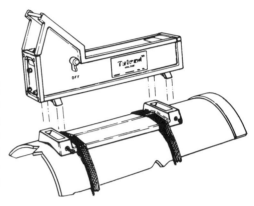

The Telrad sight uses three lighted target rings against the "real sky" allowing you to see the sky the way the star charts show it (not a small upside-down, magnified portion of the sky) and making it easier to point your telescope to exactly the right spot. The Telrad is 8 inches long, weights 11 ounces and mounts on any telescope without drilling holes.

Texas Nautical Repair Company

3110 S. Shepherd
Houston, TX 77098
Phone: (713) 529-3551
Fax: (713) 529-3108
E-mail: lsstnr@neosoft.com
Internet: www.neosoft.com/~lsstnr

Product(s): Binoculars, Telescopes

Texas Nautical Repair Company is an importer and distributor of Miyauchi astronomical binoculars and Takahashi telescopes. Takahashi manufactures high quality amateur and small institution computer-controlled refractors, hyperbolic astrographs, and Cassegrain optical systems.

Photo: Texas Nautical Repair Company

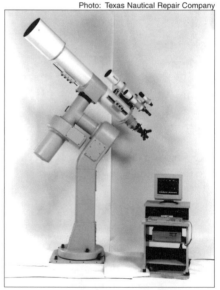

The FCT-200 8-inch apochromatic refractor and computer controlled mount is one system offered by Texas Nautical Repair Company. The entire system weighs 1 metric ton.

Torus Precision Optics, Inc.

P.O. Box 3387
Iowa City, IA 52244-3387
Phone: (319) 354-1410
Fax: (319) 358-2467
E-mail: torusoptics@msn.com
Internet: www.torusoptics.com

Product(s): Telescopes, Telescope Accessories

Torus Precision Optics currently manufactures the 16" Torus CC04 Cassegrain telescope and will introduce a 20" and 24" version of the CC04 during the later part of 1997. The Torus CC04 is a complete turnkey research telescope system. Everything needed to operate and control the telescope, including the computer, is tested at Torus' facility prior to shipment. Prices for the 16", 20" and 24" telescopes are $45,000, $75,000 and $110,000 respectively. Delivery time on the CC04 is 3 months. Larger instruments require 6-9 months.

Torus Precision Optics also offers a line of medium and large format CCD cameras featuring Kodak front illuminated and back illuminated chips (512x512 and 1Kx1K).

Typographica Publishing

313 Raphael Avenue
Middlesex, NJ 08846-1224
Phone: (732) 469-7752

Product(s): Magazines, Music

A multifaceted company, Typographica Publishing first introduced their quarterly magazine, *The Practical Observer* (TPO), back in 1985. TPO's editorial focus is primarily on techniques that serious amateur astronomers can use to study the universe, from astrophotography to more exotic methods such as radio astronomy, astromicroscopy, and micrometeorite collection. A one year subscription currently cost $12 ($13 in Canada, $15 in Mexico).

Another facet of Typograhica Publishing is Lyra Recording, which supplies music to the planetarium industry. Lyra collects demo tapes from interested artists and assembles them into a sampler cassette which they sell

The Practical Observer, published quarterly by Typographica Publishing

to planetariums with reuse rights. They also provide cassettes and CDs for resale in gift shops (contact publisher Gordon Bond for more information).

UFO Central Home Video

2321 Abbot Kinney Blvd., #200
Venice, CA 90291
Phone: (310) 578-5300
Fax: (310) 578-5308
E-mail: UFOCENTRAL@aol.com

Product(s): Videos

UFO Central Home Video has conducted a painstaking search for every UFO video ever made and as a result, they now offer an extensive collection (hundreds of titles) of fully produced video presentations on UFO's and related sciences. Topics include extraterrestrial archeology, flying saucer technology, NASA and UFO's, crop circles, sacred geometry, hyperdimensional physics, consciousness, new world order, and more. Software and audio programs can also be purchased and an extensive, illustrated catalog is available.

FLYING SAUCERS ARE REAL!

Eyewitness testimony from US Army personnel who actually recovered and transported crashed Saucers and Alien bodies.

Nuclear Physicist Stanton T. Friedman tells the complete story of the government's secret recovery of two crashed saucers with alien bodies near Roswell, New Mexico in 1947.

The story that launched "Alien Autopsy: Fact or Fiction?" on FOX-TV.™

Volume 1

Videotaped at Spaceport USA, Cape Canaveral, Florida

Unitron, Inc.

170 Wilbur Place
Mail:
P.O. Box 469
Bohemia, NY 11716-0469
Phone: (516) 589-6666
Fax: (516) 589-6975
E-mail: unitron@ix.netcom.com

Product(s): Binoculars, Specialty Optics, Telescopes

For outstanding celestial and terrestrial viewing, Unitron offers a wide variety of refractor telescopes to choose from. Models range from the 2.4" Equatorial through the 3" and 4" models. Unitron telescopes offer the amateur stargazer a first quality instrument built to exacting standards.

Photo: Unitron, Inc.

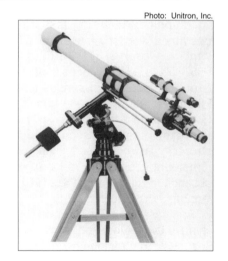

VERNONscope & Co.

5 Ithaca Road
Candor, New York 13743
Phone: (607) 659-7000
Fax: (607) 659-4000

Product(s): Spotting Scopes, Telescope Accessories

VERNONscope manufactures high precision BRANDON® eyepieces, DAKIN® Barlows, mounted colored filters, star diagonals, and other telescope accessories. The company also manufactures an 80mm f/5.6 Apochromatic spotting scope for birding and low power astronomy. Product flyers are available for $1.00.

Places of Interest

An alphabetical listing of museums, observatories, planetariums, NASA sites, and other locations of interest. For a state by state index, refer to the Places of Interest *QuickFind Index that follows this section.*

Abrams Planetarium

Michigan State University
East Lansing, MI 48824
Business Office: (517) 355-4676
Program Info: (517) 355-4672
Sky Info: (517) 332-6381
Fax: (517) 432-3838
Internet: www.pa.msu.edu/abrams/

Abrams Planetarium serves as an astronomy and space science education resource center for the mid-Michigan area. The building houses a 150-seat Sky Theater, featuring a Digistar computer graphics projector, an exhibit hall, blacklight gallery, and gift counter. Programs open to the general public are presented Friday and Saturday evenings at 8:00 p.m. and Sundays at 4:00 p.m. A family-oriented show is presented at 2:30 p.m. Sundays.

Admission to all public shows is $3.00 for adults, $2.50 for students and seniors, and $2.00 for children 12 and under. School programs are available at a reduced rate Tuesday through Thursday by reservation only. The planetarium office is open weekdays 8:30 a.m. to noon, and 1:00 p.m. to 4:30 p.m.

Adler Planetarium & Astronomy Museum

1300 South Lake Shore Drive
Chicago, IL 60605
Phone: (312) 322-0329
Fax: (312) 322-2257
TTY: (312) 322-0995
E-mail:
Elaine_Barreca@orbit.adler.uchicago.edu
Internet: astro.uchicago.edu/adler/

One of the most impressive planetariums you'll find, Adler Planetarium and Astronomy Museum also has the distinction of being the first planetarium in the Western Hemisphere. In 1923, Dr. Walther Bauersfeld, the scientific director of the firm of Carl Zeiss in Jena, Germany, developed optics that would help recreate the night sky by projecting onto the surface of a domed ceiling, giving rise to the modern planetarium. After the invention of these optics, Max Adler, a Chicago businessman, traveled to Germany to see the planetarium for himself. He was so impressed, he quickly obligated $500,000 for the construction of a planetarium in Chicago. The Adler Planetarium was opened to the public on May 12, 1930.

Sky shows include *Seeing the Invisible Universe* (50 minutes), *Comets Are Coming* (50 minutes), *The Sky Tonight* (45 minutes), *Stargazing With Meteor Mouse* for families with young children (45 minutes), *African Skies*

(50 minutes), *Solar System Vacation* (50 minutes), plus three holiday sky shows to celebrate the Christmas season. Following the 8 p.m. sky show on Fridays, live telescope images of celestial objects are transmitted to the theater from Adler's Doane Observatory. The Doane Observatory, located just east of the main planetarium building, is open for tours at scheduled times. Ask at the box office for details.

Photo: Adler Planetarium

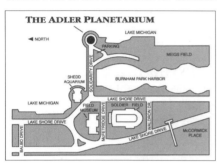

Sky show tickets are $2 per show for all ages. Admission to the museum is $3 for adults, $2 for children (age 17 and under), $2 for Senior Citizens age 65 and over. Tuesdays are FREE. Hours are Monday through Thursday, 9 a.m. to 5 p.m.; Friday, 9 a.m. to 9 p.m.; and Saturday & Sunday from 9 a.m. to 6 p.m.

Air Force Academy Planetarium

Center for Educational Multimedia
34ES/CEMM
2120 Cadet Drive USAF Academy
Colorado Springs, CO 80840
Phone: (719) 333-2779
Recording: (719) 333-2778
Fax: (719) 333-4281
E-mail: schmidtmd.34edg@usafa.af.mil

The mission of the Academy Center for Educational Multimedia (planetarium) is to support cadet training and instruction. Because of this, it may be necessary to cancel (without notice) and reschedule any public or school presentation so affected. The planetarium supports cadet instruction in a wide variety of academic and military topics, including aeronautics, astronomy, astronautics, aviation science, physics, military arts and sci-

The planetarium features an Evans & Sutherland Digistar Projector and seating for 149 persons.

ences, survival training, engineering and mathematics. The planetarium features an Evans & Sutherland Digistar Projector and seating for 149 persons. Handicap accessible. Office hours 7:30 a.m. to 4:30 p.m., Monday through Friday.

All public shows are free and sometimes include an open house at the Academy Observatory. Schools and organizations may schedule presentations for Tuesday and Thursday mornings. Groups must be 3rd grade and above. Groups of 25 or smaller must be willing to join with another group. A wide variety of presentation topics are available. Call for details and reservations.

Air Force Space and Missile Museum

Cape Canaveral Air Station
191 Museum Circle
Patrick AFB, FL 32925-2535
Phone: (407) 853-9171
Recording: (407) 853-3245
Fax: (407) 853-9172

Located at Launch Complex 26, the site of the first successful launch of an American satellite, Explorer I, in 1958, the Air Force Space and Missile Museum bills itself as "the Kitty Hawk of the Space Age," a moniker justly earned when you consider the history of the site. Complex 26 hosted 36 launches between 1957 and 1963, including the launches of "astro chimps" Gordo, Able and Miss Baker in 1958 and 1959.

The museum includes the original blockhouse (a scant 400 feet from the twin launch pads), an Exhibit Hall, and an extensive outdoor Rocket Garden that features more than 55 rockets, missiles, re-entry vehicles, and other space hardware. Highlights at the museum are numerous and include a full-scale model of Explorer I in the blockhouse, one of only two complete German V-2 engines in the United States, and Gemini II, both located in the Exhibit Hall, and a host of well-known rockets and missiles in the Rocket Garden (like Titan, Thor, Bomarc, Mace, Snark, Navaho, and others).

The museum is open to the public Monday through Friday from 10 a.m. to 2 p.m., and on weekends from 10 a.m. to 4 p.m. The staff suggests that you call before visiting, as the museum is periodically closed during fueling and other pre-launch operations at Cape Canaveral Air Station.

Photo: Air Force Space & Missile Museum

The Rocket Garden at the Air Force Space and Missile Museum

Allentown School District Planetarium

Gary A. Becker, Director
Dieruff High School
815 North Irving Street
Allentown, PA 18103-1894
Phone: (610) 820-2204
Fax: (610) 282-2497
E-mail: gabecker@itw.com
Internet: asd.planetarium/org
Internet: www.astronomy.org

The ASD Planetarium, which was founded in 1964, offers programming for school, church, and Scouting groups, senior citizens' leagues, civic organizations, even birthday parties. Day and evening programs are available, Monday through Friday, and can be arranged by calling the planetarium program office. The charge is $50 per hour.

Ames Research Center

Educator Resource Center
Mail Stop T253-2
Moffett Field, CA 94035-1000
Phone: (650) 604-3574
Fax: (650) 604-3445

NASA's Ames Research Center was founded December 20, 1939 as an aircraft research laboratory by the National Advisory Committee on Aeronautics (NACA) and in 1958 became part of National Aeronautics and Space Administration (NASA).

The Educator Resource Center is located at Ames and serves educators in the western states (Alaska, Arizona, California, Hawaii, Idaho, Montana, Nevada, Oregon, Utah, Washington, Wyoming) and trust territories of the Pacific Islands. Educators can use NASA resources at the Educator Resource Center to develop their own educational programs, gather ideas, do research, and duplicate audiovisual materials. Educator Resource Center materials reflect NASA research and technology development in such curriculum areas as life science, physical science, astronomy, energy, earth resources, the environment, mathematics, and careers in aerospace.

Teachers in disciplines other than science and mathematics are also encouraged to visit the ERC and explore ways in which aerospace materials may be incorporated into their lessons.

Educational materials available at the Center include: NASA publications, video programs (over 200 to choose from), 35mm slides, curriculum materials, lesson plans, Apple IIe and Mac software, and audio cassettes. Equipment is available to preview and duplicate slides, videotapes and audiotapes. Educators need to provide their own blank videotapes, audio tapes and computer disks, as well as slide film (Kodachrome 64, Ektachrome 100 or Ektachrome 200). Duplication equipment is available by reservation, please call (650) 604-3574. The Educator Resource Center is open Tuesday through Friday, 9 a.m. to 5 p.m. (closed for lunch from noon to 1 p.m.), and on Saturday from 8 a.m. to 3 p.m.

Bakersfield College Planetarium

1801 Panorama Drive
Bakersfield, CA 93305
E-mail: strobel@lightspeed.net

Shows at Bakersfield College Planetarium are currently geared for school groups only (mostly elementary school groups). Plans for the future include evening shows for the general public. Nick Strobel, director of the planetarium, has an award-winning web site on introductory astronomy (http://bizweb.lightspeed.net/~astronomy). Almost an "on-line textbook," Mr. Strobel uses the site in his introductory astronomy class and other astronomy and science teachers have found the site helpful in their classes. Check it out!

Nick Strobel, director of the planetarium, has an award-winning web site on introductory astronomy

Barlow Planetarium

University of Wisconsin Fox Valley
1478 Midway Rd.
Menasha, WI 54952
Sky Show Information: (414) 832-2848
Reservations: (414) 832-2600
Fax: (414) 832-2674
E-mail: tfrantz@uwc.edu

Opening in December of 1997, the Barlow Planetarium will be the largest and most modern planetarium in the state of Wisconsin. It features a Digistar II star projector, 48-foot dome, interactive audience response seating, and a Dolby THX surround sound system to deliver an "out of this world" experience. Located in the Fox Cities of Northeastern Wisconsin, the Barlow Planetarium is easily accessible from major highways.

The Barlow Planetarium presents both topical and seasonal shows for the general public. Organized groups and schools can take advantage of special shows, times, and group rates if they schedule in advance. Most shows run one hour. Visitors are encouraged to call (414) 832-2848 for current show times and pricing information.

Opening in December of 1997, the Barlow Planetarium will be the largest and most modern planetarium in the state of Wisconsin.

Berkeley County Planetarium

Elizabeth Wasiluk, Director
Hedgesville High School
Rt. 1 - Box 89
Hedgesville, WV 25427
Phone: (304) 754-3354
Fax: (304) 754-7445

Photo: Berkeley County Planetarium

The Berkeley County Planetarium is open by reservation only, and for groups of 25 or fewer. There is no charge. All programs are interactive, meaning visitors will take part in each program they see.

Planetarium Director Elizabeth Wasiluk readies her Spitz 373 for another interactive planetarium program at Berkeley County Planetarium.

Bishop Planetarium

South Florida Museum
201 10th Street West
Bradenton, FL 34205
Phone: (941) 746-4131
Internet: www.manatee-cc.com/planets

Opened in 1966, the Bishop Planetarium features 220 seats under a 50-foot diameter dome with a Spitz STP Star Projector. Video projection, lasers, slides, and special effects combine with the projector to produce a variety of shows. Weekday mornings at the planetarium are reserved for the school groups that bring thousands of students to the facility each year. Afternoon shows for the general public are held at 1:00, 2:30, and 4:00 p.m. Evening shows are held on Friday and Saturday at 7:30 and 8:30 p.m., followed by laser shows at 9:00 and 10:30 p.m. The planetarium offers a family program on Saturday mornings at 10:30 a.m. This program is followed by hands-on activities in the museum. The South Florida Museum also has a rooftop observatory which is open, weather permitting, every Friday and Saturday evening for celestial observing, and on Saturday afternoons for solar observing. Reservations for schools or adult groups can be made by calling (941) 746-4132 ext. 13.

The South Florida Museum also has a rooftop observatory

Brackbill Planetarium

Eastern Mennonite University
1200 Park Road
Harrisonburg, VA 22801
Phone: (540) 432-4400
Fax: (540) 432-4488
E-mail: mastj@emu.edu
Internet: beta77.emu.edu/mastjw.html

The M. T. Brackbill Planetarium was built in 1967 and uses a Spitz A-4 projector to simulate the heavens.

Service programs are offered to the public during the academic year at no charge. The primary programs are for K-12 students from the surrounding area. Over 5000 people visit the planetarium each year. There is also a natural science museum with over 6000 items on display which is part of a standard planetarium visit.

Buehler Planetarium

Broward Community College
Central Campus
3501 S.W. Davie Road
Davie, FL 33314
School Programs: (954) 475-6681
Show Information: (954) 475-6680
Sky Information: (954) 475-6734

A wide variety of entertaining and educational programs for all ages are available at the Buehler Planetarium, including school programming, science lectures, star shows, laser shows, Saturday science classes, a mobile planetarium program, and a public observatory. Thanks to a generous gift from the Emil Buehler Trust,

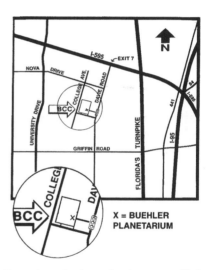

the planetarium features a Zeiss M1015 star projector and the latest audio/visual technology.

Buhl Planetarium and Observatory

Carnegie Science Center
One Allegheny Avenue
Pittsburgh, PA 15212-5850
Phone: (412) 237-3400
Internet: www.csc.clpgh.org

Sponsored by the Buhl Foundation, the Henry Buhl Jr. Planetarium and Observatory at Carnegie Science Center is a recognized leader in education and has achieved a world-renowned reputation by introducing a new model for planetarium design and technology. Combining their 3-D Digistar projections with video graphics, computer animation, interactive surround-sound, and a real-time audience participation system, the planetarium has succeeded in creating a three-dimensional "edutainment" vehicle that generates state-of-the-art scientific simulations in a highly interactive manner. The planetarium's 50-foot domed chamber has 150 reclining seats with individual flight controllers at each seat. The curriculum goes beyond astronomy to include presentations on biology, art, drama, navigation, geometry, mythology, and technology. The Carnegie

> *Carnegie Science Center is a recognized leader in education and has achieved a world-renowned reputation*

Science Center itself plays host to hundreds of thousands of visitors each year—an estimated 700,000 in 1996 alone—and opens daily at 10 a.m. (closing times vary).

Burke Baker Planetarium

Dr. Carolyn Sumners, Director
Houston Museum of Natural Science
One Hermann Circle Drive
Houston, TX 77030
Phone: (713) 639-4630
Fax: (713) 639-4635
E-mail: planetarium@hmns.mus.tx.us
Internet: hmns.org

The Burke Baker Planetarium is the largest planetarium in Texas. Opened in 1964, it features a 50-foot dome and seating for 218 people. The Digistar star projector, along with state-of-the-art laser and automation systems, combine to provide exciting educational astronomy shows and entertaining laser concerts. Almost one quarter million people pass through the planetarium each year; approximately one-half of these visitors are elementary and secondary-school groups. Astronomy shows and matinee laser shows are presented daily ($3.00/adult, with discounts for children, seniors, large groups and school groups). On Friday and Saturday nights, the planetarium shows rock laser concerts ($6.00/person). The Houston Museum of Natural Science also operates The George Observatory, a satellite facility in Brazos Bend State Park; the observatory opens to the public every Saturday night.

> *the largest planetarium in Texas*

California State University, Northridge Planetarium

Department of Physics & Astronomy
Northridge, CA 91330
E-mail: physics@galileo.csun.edu
Internet: davinci.csun.edu/~astro/
plnt.html

The California State University, Northridge Planetarium features a 40-foot dome, 105 seats, a Spitz 512 system, and a wide range of audio and video support systems. Video is provided by an Electrohome video projection system that produces a video image about 25 feet along the diagonal. The planetarium is heavily used during most parts of the day for lower division astronomy classes, but can be made available to nearby school groups in the late afternoon (i.e. after 3 p.m. Pacific time). The planetarium is also used by the Los Angeles School District's Granada Hills High School Magnet School program.

Carl Sandburg Middle School Planetarium

Drexell George, Planetarium Teacher
8428 Fort Hunt Road
Alexandria, VA 22308
Phone: (703) 799-6169

The Carl Sandburg Middle School planetarium has a Spitz A4 planetarium projector, a 30-foot diameter dome, and seating for 76.

Casper Planetarium

Susan Peterson, Director
904 N. Poplar St.
Casper, WY 82601
Phone: (307) 577-0310
E-mail: speters@trib.com
Internet: www.trib.com/WYOMING/
NCSD/planetarium.html

Casper Planetarium offers public programs throughout the year

The Casper Planetarium was built in 1966 by the Natrona County School District. It has been in continuous operation since the 1966-67 school year providing an outstanding educational experience for students of the Natrona County schools. Many schools and colleges throughout Wyoming also use the planetarium for classes in conjunction with their curriculum.

The Casper Planetarium offers public programs throughout the year, providing audiences with a unique multi-media encounter with space and astronomy-related subjects. Programs range from the latest frontiers of astronomical research to astronomy's earliest beginnings. The Casper Planetarium also maintains a gift shop specializing in astronomy and science-related toys, posters, books, t-shirts, star charts and games.

Cernan Earth and Space Center

Triton College
2000 N. Fifth Avenue
River Grove, IL 60171
Office: (708) 456-0300 ext. 3372
Program Line: (708) 583-3100
Fax: (708) 583-3121
E-mail: cernan@triton.cc.il.us
Internet: www.triton.cc.il.us/cernan/ cernan_home.html

Included among the exhibits are Captain Cernan's Apollo 10 spacesuit and coverall garment

Named for astronaut Eugene A. Cernan, who flew aboard the Gemini 9, Apollo 10, and Apollo 17 space missions, the Cernan Earth and Space Center features a 100-seat dome theater with a 44-foot diameter. The theater presents earth and sky shows (including C-360 wraparound films), children's shows, and laser light shows. The Center has a Space Hall which contains a variety of exhibits related to astronomy, space science and earth science. Included among the exhibits are Captain Cernan's Apollo 10 spacesuit and coverall garment, an Apollo lunar landing diorama, an interactive moon phase exhibit, a color weather radar station, a 1/15th scale Space Shuttle model, an Illinois fossil exhibit, a telescope exhibit, a current events bulletin board, Apollo 17 moon gloves and other Apollo artifacts. Outdoor exhibits include an Apollo practice capsule and a Nike Tomahawk missile. The Center is open seven days a week. Call for hours and show times.

Photo: Cernan Earth and Space Center

Chabot Observatory and Science Center

4917 Mountain Blvd.
Oakland, CA 94619-3014
Phone: (510) 530-3480
Fax: (510) 530-3499
E-mail: cosc@cosc.org
Internet: www.cosc.org

The Chabot Observatory and Science Center traces its roots back to 1883 when Antoine (Anthony) Chabot presented the Oakland School District with an 8-inch refractor that gave birth to the Oakland Observatory. Following Mr. Chabot's death in 1888, the city of Oakland renamed the facility in his honor. A wide variety of programs and classes are available at the Chabot Observatory and Science Center, including classes tailored for children (6-12), families, and adults.

On October 18, 1996, after ten years of planning, the groundbreaking event for a new $52 million Science Center took place. Completion of the new, advanced facility is scheduled for the fall of 1999. The Science Center will feature a Carl Zeiss Jena Universarium Model VIII fiber optic planetarium projector system - the first to be installed in North America.

THE NEW CHABOT OBSERVATORY & SCIENCE CENTER

Chaffee Planetarium and Veen Observatory

Dave DeBruyn, Planetarium Curator
Van Andel Museum Center
272 Pearl N.W.
Grand Rapids, MI 49504-5371
Phone: (616) 456-DOME (456-3663)
Fax: (616) 456-3873
E-mail: staff@grmuseum.org
Internet: www.grmuseum.org

Housed in Van Andel Museum Center of the Public Museum of Grand Rapids, Roger B. Chaffee Planetarium offers spectacular sky shows in its state-of-the-art theater. The planetarium features 145 uni-directional seats for comfortable viewing, preprogrammed audio-visual shows for maximum dramatic impact, multiple dynamic speakers as well as a Digistar simulation device capable of projecting a view of the universe from any point in space. Chaffee Planetarium is suitable for a variety of programs, such as concerts and dramas using special sound and visual effects in addition to sky shows.

Admission price for sky shows is $1.50 in addition to museum admission. Evening Laser Light Shows are $5.00. Show times vary throughout the year. The Planetarium, named after the Grand Rapids born astronaut who lost his life in the 1967 Apollo spacecraft fire, is visited by over 90,000 people each year.

Photo: Chaffee Planetarium

The James C. Veen Observatory, operated by the Grand Rapids Amateur Astronomical Association, offers observing sessions for the public on the second and last Saturday evening of the month, April through October, beginning at dusk. There is a $1.50 observatory admission fee for adults, $.50 for children under 12. Call (616) 897-7065 for information.

Challenger Learning Center

Gayle F. Glusman, Flight Director
Wanda Williams, Mission Commander
The Louisiana Arts & Science Center
100 South River Road
Baton Rouge, LA 70802
Phone: (504) 344-5272
Fax: (504) 344-9477
E-mail: lascclc@aol.com

The Challenger Learning Center is an exciting, hands-on facility that features a Mission Control room designed after NASA's Johnson Space Center and a space lab simulator designed to give simulation participants the feeling of working in a Space Lab. The center offers three simulated space missions; *Rendezvous with Comet Halley* (grades 5-8), *Return to the Moon* (grades 6-9), and *Encounter Earth* (grades 8-12).

School programs are presented Monday through Friday by reservation only. Teachers wishing to have their class participate must attend an "in-service" during the summer. The cost of the in-service, all curriculum material, and the student flight is $275.00.

In this dynamic environment, students use principals of science, mathematics, and technology to complete mission tasks. Embedded throughout the simulations are opportunities for students to use multiple process skills, including manipulation, procedural, and critical thinking skills. The Challenger Learning Center provides students with an opportunity to apply the skills they have learned in the classroom. The experience creates a co-operative learning atmosphere under-scored by teamwork, communication, problem solving, and decision-making.

"Flights" open to the general public are offered on the weekends by reservation only. These flights are conducted only in the Space Lab simulator. A group of 14-16 is needed for each flight. The cost is $15 per person. Members of the Louisiana Arts and Science Center can participate for only $10 per person.

Challenger Learning Center

Kansas City Museum
3218 Gladstone Blvd.
Kansas City, MO 64123-9989
Phone: (816) 483-8326
Internet: www.kcmuseum.com

The Challenger Learning Center at Kansas City Museum offers teach-ers and students the opportunity to learn more about space and science through a space flight simulation experience that lasts about 2 hours 15 minutes. "Flights" for families, which last about 45 minutes, are also offered, as are adult and corporate team-building missions.

Chemeketa Community College Planetarium

Tom McDonough, director
Chemeketa Community College
4000 Lancaster Dr. NE
P.O. Box 14007
Salem, OR 97309
Phone: (503) 399-5161
Recording: (503) 399-5200
School Reservations: (503) 399-5246
Fax: (503) 399-5214
E-mail: mcdt@chemek.cc.or.us
Internet: www.chemek.cc.or.us

Saturday afternoon programs for the family are scheduled periodically, with the theme generally changing from week to week. These programs are also suitable for clubs and Scout groups.

The Chemeketa Community College Planetarium consists of a 35-foot dome raised above 60 comfortable seats which lean back for easy viewing. Located on the college campus, the planetarium serves the general public, school groups, and the college. School groups and other organizations may schedule workshops and presentations in the planetarium on weekdays. Shows may emphasize elements of the solar system or stellar system to correspond with classroom studies. A fee of $30 is charged for groups. Saturday afternoon programs for the family are scheduled periodically, with the theme generally changing from week to week. These programs are also suitable for clubs and Scout groups; a schedule may be obtained by calling the planetarium office. Programs for the general public are presented at 7:30 p.m. Friday nights during the academic year. Ticket prices for the Saturday afternoon and Friday night programs are $2 for adults and $1 for children, with a $5 family rate.

Chesapeake Planetarium

Robert J. Hitt Jr., Director
300 Cedar Road
Chesapeake, VA 23320
Phone: (757) 547-0153 ex. 208

Public programs are presented on Thursday nights at 8 p.m.

The Chesapeake Planetarium was the first planetarium built by a public school system in the state of Virginia. It was constructed with funds provided by the National Defense Education Act of 1958 (NDEA). Each year the planetarium presents programs for grades K-12 as well as public programs during the evening hours. Programs are adapted to grade levels and provide a general overview of space science and astronomy. Public programs are presented on Thursday nights at 8 p.m., and there is a telescope observing session after each program. Reservations are required. The planetarium office is open weekdays 8 a.m. to 4 p.m.

Christa McAuliffe Planetarium

3 Institute Drive
Concord, NH 03301
Phone: (603) 271-7831
Show Information: (603) 271-STAR
Fax: (603) 271-7832
Internet: webster.state.nh.us/cmp/ homepage.html

The Christa McAuliffe Planetarium has seating for 92 and offers public shows on weekends and midweek afternoons. Free monthly SkyWatch observing and lecture events are also offered and an Astronomy Day Celebration is held each May. The planetarium is closed on Mondays, major holidays and the last two weeks in September. Admission is $6 for adults and $3 for children and senior citizens. Group rates and special rentals are available. Though walk-ins are welcome, the planetarium recommends purchasing tickets in advance.

College of St. Catherine Observatory

Department of Physics
College of St. Catherine
2004 Randolph Avenue
St. Paul, MN 55105
E-mail: askastro@stkate.edu
Internet: www.stkate.edu/physics

open to the public one or two nights per week during the academic year and on special occasions

The observatory at the College of St. Catherine is open to the public one or two nights per week during the academic year and on special occasions. Equipment includes a 14" Celestron Compustar with SBIG7 CDD camera. The site plays host to the NSF Chautauqua Short Course for college faculty: "Using New Technologies for Teaching Astronomy." The observatory will also host the "Rocket Science Leadership Course" for selected teachers in Minnesota as part of "Partners in Aerospace." Launch date for this program is May 1998.

Community College of Southern Nevada Planetarium

Dr. Dale A. Etheridge, Director
3200 E. Cheyenne Ave.
North Las Vegas, NV 89030
Phone: (702) 651-4SKY
Fax: (702) 643-6428
E-mail: drdale@nevada.edu
Internet: www.ccsn.nevada.edu/
Planetarium/

The Planetarium at the Community College of Southern Nevada offers public presentations every Friday and Saturday. Programs consist of a double feature multi-media planetarium presentation and a hemispheric motion picture. In addition to the planetarium theater, The Planetarium has a small gift shop and an observatory with telescopes ranging from 6" to 16" in aperture. The site plays host to the Nevada NASA Regional Teacher Resource Center and is a Project Astro site working in cooperation with the Astronomical Society of the Pacific. After the last public show each Friday and Saturday evening the Student Observatory is open for direct telescopic viewing of celestial objects.

Admission prices for public presentations are $3.50 general admission and $2.25 for seniors, children under 12 and CCSN students. Presentation times are 6:00 p.m. and 7:30 p.m. on Fridays and 3:30 p.m. and 7:30 p.m. on Saturdays. Telescope observing sessions at 8:30 p.m. on Fridays and Saturdays (weather permitting) are free.

The Planetarium has a small gift shop and an observatory with telescopes ranging from 6" to 16" in aperture

The Connecticut River Valley Astronomers' Conjunction

Richard Sanderson
P.O. Box 2793
Springfield, MA 01101-2793
Phone: (413) 782-7054
E-mail: rsanderson@juno.com

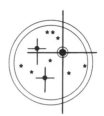

Now in its 15th year, the Astronomers' Conjunction attracts astronomy enthusiasts from across New England and New York for a weekend of education, observing, and fellowship in a beautiful rural setting. This annual weekend event features an interesting and varied selection of astronomy papers presented by professional and amateur astronomers. Other activities include a swap table, solar and night-time observing sessions, an outdoor barbecue dinner, field trips, and beautiful camp sites at nearby Barton Cove on the Connecticut River.

a weekend of education, observing, and fellowship

The Conjunction is held at the Northfield Mountain Recreation & Environmental Center in Northfield, MA, which also offers self-guided nature hikes along its miles of trails as well as riverboat rides on the Connecticut River.

The Custer Institute

P.O. Box 1204
Main Bayview Road
Southold, NY 11971
Phone: (516) 765-2626
BBS: (516) 765-2782
Internet:
ourworld.compuserve.com/
homepages/custer

Located on Main Bayview Road off Route 25, The Custer Institute is open to the public every Saturday night (weather permitting). Named after the grandniece of General George Armstrong Custer, the institute's collection of astronomical equipment includes two antique Alvan Clark refractor telescopes (a 5-inch and a 6-inch). The institute currently has about 100 members and publishes The Custer Comment, *a monthly newsletter.*

Davis Planetarium

Maryland Science Center
601 Light Street
Baltimore, MD 21230
Phone: (410) 685-2370
Fax: (410) 545-5974
Information: (410) 685-5225
Reservations: (410) 545-5929
Starline: (410) 545-5918
Internet: www.mdsci.org

sponsors Camp-Ins for school groups, grades four through six, where students spend the night at the Center

The Maryland Science Center features three floors of hands-on exhibits, a science discovery room for children aged four through seven, an IMAX Theater and the Davis Planetarium. The Center is in the planning stages to re-open their observatory, which features an Alvin Clark telescope. The Center is also affiliated with the Baltimore Astronomical Society (BAS), which meets at the Center on the second Tuesday of each month.

The Davis Planetarium seats 140 under a 50-foot dome illuminated by 150 projectors that display 8,500 stars, rotating planets, and exploding stars.

A selection of school programs is available for pre-K students through grade 12. The Science Center also sponsors Camp-Ins for school groups, grades four through six, where students spend the night at the Center and participate in hands-on science workshops, view a planetarium show, see an IMAX film, and explore three floors of exhibits. Call the Supervisor of Camp-In Programs at (410) 545-5955 for additional information.

Planetarium shows are available daily. Science Center admission is $9 for adults, $7 for children, students, military, and senior citizens.

Davis Planetarium

Richard S. Knapp, Director
201 E. Pascagoula Street
Mail:
P. O. Box 22826
Jackson, MS 39225-2826
Phone: (601) 960-1550
Show Information: (601) 960-1540
Fax: (601) 960-1555
Internet: www.msssp.org (Mississippi Student Space Station™)

offers two-week summer residencies in science education using a Mission Control center and Athena, its Student Space Station™

The Russell C. Davis Planetarium's McNair Space Theater features an 18.2-meter dome, seating for 190, a Minolta Series IV star projector, lobby exhibits and displays, and a gift shop. Located in the heart of downtown Jackson, the planetarium serves Metro Jackson residents and visitors as well as school groups of all ages from around the state. School programs are presented Monday through Friday by reservation only. Admission is $1.50 per student and

Photo: Davis Planetarium

over 30,000 students visit the Planetarium each year. Programs open to the general public are presented Tuesday through Saturday at 7:30 p.m. with matinees at 2 and 4 p.m. on Saturday and Sunday. Admission to public shows is $4.00 for adults and $2.50 for children under 12 and senior citizens. Laser/light shows are presented Friday and Saturday at 9 and 10 p.m. Admission to laser shows is $5.00 for adults and $3.50 for children under 12 and senior citizens.

From June through August, weekday matinees are presented at noon (public show), 1 p.m. (laser show), and 2 p.m. (public show). The planetarium also offers two-week summer residencies in science education using a Mission Control center and *Athena*, its Student Space Station™ simulator. Planetarium offices are open Monday through Friday from 8 a.m. to 5 p.m.

Detwiler Planetarium

Richard R. Erickson, Director
Lycoming College
Williamsport, PA 17701-5192
Phone: (717) 321-4284
Fax: (717) 321-4090

The Detwiler Planetarium has seating for 80 in a 9.1-meter dome. In addition to college classes and campus events, the planetarium serves public and private schools throughout north-central Pennsylvania. Live "sky shows" are presented to school groups from prekindergarten up through high school age. Admission for school groups is free, but by advance reservation only. Shows can be designed for individual visiting school groups and specific topics may be requested when reservations are made.

Discovery Park

Gov Aker Observatory
1651 Discovery Park Blvd.
Safford, AZ 85546
Phone: (520) 428-6260
Fax: (520) 428-6260
E-mail:
discovery@discoverypark.com
Internet: www.discoverypark.com

Located in scenic Gila Valley, Discovery Park's Gov Aker Observatory (GAO) is a premier museum of science, complete with interactive galleries. You'll have the opportunity to take a ride on Gov Aker Observatory's full-motion flight simulator, or experience craters being blasted into the surface of the moon. The observatory also offers classes exploring topics such as the mythology of the constellations, telescope making, and rocketry. Star parties, guest speakers, and other special events are hosted at the observatory each month.

Discovery Park also offers weekend tours to Mount Graham International Observatory. These tours include an interpretive journey up Mt. Graham, lunch near the summit, a personally guided tour of the Vatican Advanced Technology Telescope and the Heinrich Hertz Submillimeter Telescope. This is followed by an evening of stargazing through the Gov Aker Observatory's 20-inch reflecting telescope.

Annual membership packages are available for those wishing to take advantage of Gov Aker Observatory all year long. Membership includes free regular admission, a one year subscription to *The Explorer*, single-use guest passes, and reduced rates on select events and programs.

Doran Planetarium

Laurentian University
Ramsey Lake Road
Sudbury, Ontario
Canada, P3E 2C6
Phone: (705) 675-1151, ext. 2227
Fax: (705) 675-4868
E-mail: plegault@nickel.laurentian.ca
Internet: alumni.laurentian.ca/www/
physics/planetarium/planetarium.html

Doran Planetarium, located in the Fraser Building on the Laurentian University campus, is available most days of the year for demonstrations in English or French for primary and secondary students and other interested groups. Programs include *Astronomy I* (Grades 1-4), *Astronomy II* (Grades 5-8), *Voyager Encounters* (narrated by Patrick Stewart), *A Search for Extraterrestrial Life*, and others.

Dreyfuss Planetarium

The Newark Museum
University Heights
49 Washington Street
Mail:
P.O. Box 540
Newark, NJ 07101-0540
Phone: (973) 596-6529
Skyline: (973) 596-6611
Internet: info.rutgers.edu/newark/
museum/dreyfuss.html

Located at the elegant Newark Museum, the Alice and Leonard Dreyfuss Planetarium was originally opened in 1953 and has undergone several upgrades since then. The planetarium currently features a 24-foot diameter perforated aluminum dome that seats 50. Public programs cover such diverse subjects as extraterrestrial life, Mars, space exploration, and Native

American sky mythology. Over 20,000 school children visit the Dreyfuss Planetarium each year to participate in introductory astronomy programs. The planetarium also has a STARLAB Outreach Education program that allows them to bring a portable planetarium called STARLAB to schools throughout New Jersey. The inflatable STARLAB is 12 feet high, 25 feet in diameter, and can accommodate 25 to 30 students at a time. The Newark Skyline offers a recorded night sky update.

Photo: Dreyfuss Planetarium

Dreyfuss Planetarium's star projector

Dryden Flight Research Center

Public Affairs Office
Edwards, CA 93523
Phone: (805) 258-3449
Tours: (805) 258-3446
**   or (805) 258-3460**
E-mail: pao@dfrc.nasa.gov

The Dryden Flight Research Center is NASA's premier installation for aeronautical flight research. It is located at Edwards AFB, California, on the western edge of the Mojave Desert 80 miles north of metropolitan Los Angeles. In addition to carrying out aeronautical research, the Center also supports the space shuttle program as a primary and backup landing site, and as a facility to test and validate design concepts and systems used in development and operation of the orbiters.

Tours of the Dryden Flight Research Center are offered free of charge Monday through Friday (except for federal holidays and shuttle landing days) at 10:15 a.m. and 1:15 p.m. on a "reservation-only" basis. The tour lasts about 90-minutes and is open to groups and individuals (students or other youth groups must have one supervising adult per five children). Each tour begins with a film about the facility. Guests are then escorted through two hangers where current research aircraft may be viewed.

For information or to schedule a tour, please call (805) 258-3446, or (805) 258-3460, or send email to: Maryann_Harness@mail.dfrc.nasa.gov.

Since the first orbital flight of the Space Shuttle in April 1981, the majority of landings have been at Dryden.

Edinboro University Planetarium

Edinboro University
Edinboro, PA 16444
E-mail address: dhurd@edinboro.edu
Internet: www.geos.edinboro.edu/
PLANET.HTML

The Edinboro University Planetarium features a 30-foot dome, 61 uni-directional seats, a Spitz A3PR projector, and an East Coast Controls Projection System. The Planetarium supports instruction in University courses, and is also host to approximately 10,000 elementary and secondary school students each year from schools located in northwest Pennsylvania, western New York, and northeast Ohio. In addition, several programs are offered during the year for the general public.

The Planetarium has developed participatory programs that meet specific grade-level needs: *Introduction to the Planetarium* (1st grade); *Can You Tell Me Which Way?* (Directions - 2nd grade); *Island at the Edge of the World* (Constellations/Mythology - 3rd grade); *Our Nearest Neighbors* (Solar System - 4th grade); *Far Out* (Stars, Nebulae, Galaxies, and more - 5th grade); and *Adapting to Space* (Space Flight, Weightlessness, Toys in Space - 6th grade). Additional programs for students in kindergarten and in junior/senior high school are also available.

Most programs provide teachers with post-visit activities that enhance the participatory program. Programs should not be limited to science classes, as they are interdisciplinary and include topics such as creative writing, mathematics, myths, and more.

Edmonton Space & Science Centre

11211 - 142 Street
Edmonton, AB T5M 4A1
Phone: (403) 452-9100
Tickets & Information: (403) 451-3344
Fax: (403) 455-5882
E-mail: essc@planet.eon.net
Internet: www.ee.ualberta.ca/essc

The Edmonton Space & Science Centre is one of Northern Alberta's premiere attractions, hosting more than 425,000 visitors annually. The stunning 57,000 square-foot facility houses the Centre's IMAX® and Margaret Zeidler Star Theatres, several exhibit galleries, the Challenger Learning Centre (the first *International* Challenger Learning Centre), an observatory, DOW Computer Lab, and an amateur ham radio station.

The Margaret Zeidler Theatre has the largest planetarium dome in Canada and uses over 200 computer-

Photo: Edmonton Space & Science Center

controlled projectors, special effects projectors, and a 2-watt krypton laser system for Music Laser Light and planetarium shows. Shows include titles such as *UFOs*, *Laser Symphony of the Stars*, and *Dr. Fantastic's Amazing*

Comet Show. After the show, don't forget to visit the Centre's public observatory, where you can take a visual journey through our solar system to view the sun, moon, planets, stars, and galaxies.

Einstein Planetarium

National Air and Space Museum
National Mall
7th and Independence Avenue, S.W.
Washington, D.C. 20560
Phone: (202) 357-1686 or 357-2700
Sky Information: (202) 357-2000
Internet: www.nasm.edu

The Albert Einstein Planetarium at the National Air and Space Museum (Gallery 201) features a 70-foot dome, a Zeiss Model Via planetarium instrument, seating for 230, video captioning, and an infra-red listening system that simultaneously translates the feature's narration into French, German, Japanese, and Spanish.

. . . at the National Air and Space Museum

El Paso ISD Planetarium

John R. Peterson, Director
Cory L. Stone, Assistant
El Paso School District
6531 Boeing Drive
El Paso, TX 79925
Info Line: (915) 779-4400
Office: (915) 779-4317
Fax: (915) 779-4098
E-mail: johnrp@tenet.edu

The El Paso Independent School District (ISD) Planetarium features a 40-foot dome, seating for 120, a Spitz 512A Planetarium Projector, ECCS Control System, over 70 astronomical, special effect, video, and laser projectors. The planetarium presents programs to all students in grades 2, 5, 6, and 8 in the El Paso District as well as over 140 programs per year for other school districts and the general public. All programs cost $2.00 per person and include a tour of the evening sky, plus a multimedia program selected from a variety of astronomical topics. To find out about school or public programs or for information on the current night sky, call 779-4400. The planetarium also contains a 2,500 sq. ft. exhibit area that is open to the public weekdays from 8 a.m. to 4:30 p.m.

over 140 programs per year for other school districts and the general public

Europlanetarium

Planetariumweg 19
3600 Genk
Belgium
Phone: 32-89-35-27-94
Fax: 32-89-36-40-50
E-mail: planetar@skynet.be

Image: Europlanetarium

The most advanced planetarium in Belgium, Europlanetarium of Genk has a dome of 12.5 meters and can seat up to 90 adults. A public observatory (the Limburgian Observatory) and exhibition rooms are collocated with the planetarium and a large auditorium and some classrooms are available. Planetarium presentations are also presented in English for groups of 25 or more. The site mainly concentrates on school groups and hosts a youth club and senior club of amateur astronomers. The planetarium currently has over 400 members.

Falls Church High School Planetarium

7521 Jaguar Trail
Falls Church, VA 22042
Phone: (703) 207-4110
Fax: (703) 207-4097
E-mail: gpurinto@pen.k12.va.us
Internet: www.fcps.k12.va.us/DIS/OHSICS/planet

The Falls Church High School Planetarium is one of nine planetaria in Fairfax County Public Schools in Virginia. The planetarium has a Spitz A3P sky projector with a 30-foot dome and 70 seats. Planetarium field trips are currently part of the fourth, fifth, and eighth grade program of studies in Fairfax County, and a variety of high school classes also use the facility.

Fels Planetarium

Franklin Institute Science Museum
222 North 20th Street
Philadelphia, PA 19103
Planetarium: (215) 448-1292
Internet: sln.fi.edu

The Fels Planetarium was the second planetarium built in the United States, and was made possible through a generous donation by Samuel S. Fels in 1933. A Digistar projection system was added to the 330-seat planetarium in 1989. The Franklin Institute Science Museum also features the Tuttleman Omniverse Theater, which projects onto a 4-story domed screen, and the Joel N. Bloom Observatory. A host of programs and exhibits are available, offering an exciting and educational outing for the entire family.

Fernbank Science Center

156 Heaton Park Drive, N.E.
Atlanta, GA 30307-1398
Phone: (404) 378-4311
Fax: (404) 370-1336
E-mail: fernbank@fernbank.edu
Internet: www.fernbank.edu

Part of the DeKalb County School System, Fernbank Science Center is a major science center that includes a 9,000 square foot exhibit hall, a 500-seat, 70-foot diameter planetarium, an observatory with a 36-inch telescope, and laboratories in various areas of science. The exhibit hall features the original Apollo 6 Command Module, Georgia's official moon rock, a dinosaur exhibit, a current meteorology exhibit, and much more. The Center is open to the public and to school groups daily, except school holidays. The observatory is open until 10:30 p.m. on clear Thursday and Friday evenings. There is an admission charge to the planetarium, which offers a variety of sky programs (closed on Mondays). All other exhibits are free.

Photo: Fernbank Science Center

Fiske Planetarium

University of Colorado
Regent Drive
Boulder, CO 80309
Phone: (303) 492-5002

Fiske Planetarium often hosts or conducts workshops on teaching astronomy and science topics for teachers of all age groups

Each year more than 25,000 people visit Fiske Planetarium, the largest planetarium between Chicago and Los Angeles. The planetarium is a place for questions and answers. What is it like to stand on the surface of Mars or go for a ride on a comet? How do rockets work? What does a meteorite feel like? Visitors discover the answers to these questions and hundreds more as they participate in star shows (the planetarium seats 210), workshops, and labs.

Fiske Planetarium is located on the University of Colorado campus on Regent Drive. Observing begins after sunset, weather permitting (excluding University Holidays). Please call to reserve observing time for groups. Fiske Planetarium often hosts or conducts workshops on teaching astronomy and science topics for teachers of all age groups. Call for information on scheduled workshops or to discuss special programs. The Fiske Planetarium theater and lobby are also available for rental and the staff will present any of the available programs, or assist you in producing your own. Call for details and prices when organizing your special reception, meeting, or presentation. The Sommers-Bausch Observatory is also located at the University of Colorado (see separate entry in *Places of Interest*).

Flandrau Science Center Planetarium & Observatory

University of Arizona
Tucson, AZ 85721
Phone: (520) 621-4515
Astronomy Hotline: (520) 621-4310
Internet: www.seds.org/flandrau

The Flandrau Science Center offers a variety of astronomy and laser shows for the public, a gift shop, hands-on displays, a public observatory, mineral museum, and more. For recorded program information and building hours, call (520) 621-STAR.

Fleischmann Planetarium

Mail Stop 272
University of Nevada, Reno
Reno, NV 89557-0010
Office/Reservations: (702) 784-4812
Recorded Schedule: (702) 784-4811
Skyline: (702) 784-1-SKY

The Fleischmann Planetarium offers a variety of fascinating shows seven days a week, such as *What if the Moon Didn't Exist?*, *Mars: Planet of Life?*, *Comet: From Ice to Fire*, and others. Sky Night lectures are provided every Monday at 7 p.m. These live 30-minute lectures examine the current night sky, identifying constellations, planets, and special astronomical objects and events visible from northern Nevada. Tuesday through Friday during the school year, the planetarium hosts visiting students from area schools and offers a selection of programs designed especially for students in grades K-12. SkyDome 8/70 feature films are available for those wishing to experience the realism created by projecting a motion picture onto the planetarium dome, surrounding the audience with sights and sounds.

The Fleischmann Planetarium also offers a public observatory that houses a 12-inch reflecting telescope (open Friday nights), astronomy classes, model rocket classes, a gift shop, and the Hall of the Solar System, an astronomical museum that features exhibits on astronomy, earth and space science, and Nevada meteorites.

Admission to the building, gift shop, Hall of the Solar System and public observatory is free. Theater ticket prices for each planetarium show and SkyDome 8/70 double feature presentation are $6 for general admission, $4 for children under 13 and seniors (60+).

Foran High School Planetarium

J. A. Foran High School
80 Foran Drive
Milford, CT 06460
Phone: (203) 783-3510
Fax: (203) 783-3635

The Foran High School Planetarium seats 70 and is open to both public and private school groups (a small admission fee is charged). The Astronomical Society of New Haven holds its monthly meetings in the planetarium and sponsors bi-annual planetarium shows for the public and scout groups. Foran High School also has an astronomy club with approximately 10 members. Linda Marks serves as the groups advisor.

Freeman Astronomy Center

Mark A. Trotter, Director
Mail:
P.O. Box 870610
New Orleans, LA 70187-0610
Phone: (504) 243-3385
Fax: (504) 242-1889
Show Information: (504) 246-STAR
E-mail: strotter@communique.net
Internet: www.communique.net/~strotter/

The Judith W. Freeman Astronomy Center at the Louisiana Nature Center features a 30-foot dome, seating for 66, a Spitz A4 projector, and laser graphics capability. The planetarium offers group programs for public and private schools, public programs on weekends, and Laser Rock concerts on Friday and Saturday nights. Admission to the group programs is $2.00 per person and is by advance registration only. Public admission is included in general admission to the Nature Center ($4 for adults, $2 for children, and $3 for senior citizens). Admission to the Laser Rock concerts is $5 per person. Current programs include *The Sky Tonight*, *Planet Patrol*, and the *Family Laser Show*.

Freeport McMoRan Daily Living Science Center Planetarium & Observatory

Michael D. Sandras, Curator
409 Williams Blvd.
Kenner, LA 70062
Phone: (504) 468-7229
Fax: (504) 468-7599
E-mail: astrox@ix.netcom.com

The Freeport McMoRan Planetarium and Observatory features a 24-foot dome with seating for 43 and an observatory with a 14" Schmidt Cassegrain telescope. This facility serves as an educational facility for local school systems and the general public. The planetarium is open by reservation Tuesday through Saturday 9 a.m. to 5 p.m., with public shows offered Tuesday through Friday at 2 p.m. and Saturdays at 10 and 11 a.m., and again at 2 and 3 p.m. Admission to the planetarium is $1.00 per person. The observatory is open on Friday and Saturday evenings from dark until 11 p.m. Admission to the observatory is $1.00 for adults and $.50 for children.

Fremont Peak Observatory

P.O. Box 787
San Juan Bautista, CA 95045
Phone: (408) 623-2465
Internet: www.astronomy-mall.com/
fpoa/

Public programs in astronomy are offered from April through October

Fremont Peak Observatory is operated by members of the Fremont Peak Observatory Association, a nonprofit organization which contracts with Fremont Peak State Park to provide astronomical education to the public. The observatory is located at an elevation of 3,000 feet in the Gavilan Mountains about 50 miles south of San Jose.

The observatory houses a 30-inch Newtonian reflecting telescope which the association has used to show objects in the night sky to tens of thousands of park visitors. Public programs in astronomy are offered from April through October and consist of a lecture followed by viewing through the 30-inch and other telescopes that may be set up outside the building. Information on public programs and membership is available by mail, telephone or from Fremont Peak's web site.

Gale Observatory

Physics Department
Grinnell College
Grinnell, IA 50112
Phone: (515) 269-3016
Fax: (515) 269-4285

Located at a relatively dark site on the north edge of the Grinnell College campus, the Grant O. Gale Observatory is used primarily for the educational and research programs of the college, but public viewing both by arranged groups and by the general public is available. The public nights are scheduled irregularly as access to the observatory permits. The observatory's primary instrument is a 24-inch f/13.5 Cassegrain reflector with full computer control, both from the dome and from an adjacent control room. Research programs at the college focus on the behavior of semiregular variable stars.

Gayle Planetarium

Rick L. Evans, Director
Troy State University Montgomery
Montgomery, AL 36110
Phone: (334) 241-4799
Fax: (334) 240-4309
E-mail: revans@tsurn.edu
Internet: www.tsum.edu

The W.A. Gayle Planetarium is one of the larger planetariums in the southeast. It has a star chamber with 235 seats, designed especially for simulating the natural sky by projecting images of the sun, moon, planets, stars and other celestial objects on a 50-foot domed ceiling. The versatile Spitz Space Transit instrument can project the positions of over 5,000 stars and the Milky Way as it would appear from any place on Earth at any time in the past, present, or future. As you pass through the entrance hallway to the star chamber, you will go on a journey through time. Ultraviolet lighting adds vivid depth and realism to an extensive mural illustrating the history of astronomy and the space age. This beautiful mural is unique to the Gayle Planetarium and was painted by Larry Godwin.

Approximately 25,000 school students (grades K-12) visit the planetarium each year. Programs are offered to the public Monday thru Thursday at 3 p.m. and the first and third Sundays of every month at 2 p.m. The planetarium is handicap accessible.

The Gayle Planetarium is located in beautiful Oak Park, downtown Montgomery, less than a mile from Interstate I-85.

Gibbes Planetarium

Columbia Museum of Art
1112 Bull Street
Columbia, SC 29201
Phone: (803) 799-2810
Fax: (803) 343-2150
Internet: www.scsn.net/users/planet

Founded in 1959, the Gibbes Planetarium features a 26-foot dome, Minolta MS-10 star projector, and seating for 50. A variety of programs are available and group reservations can be made for Tuesday through Friday, from 9:30 a.m. until 1:30 p.m. (minimum group size of 15, groups of 15-30 can expect to be combined with another group). Call for details.

Gladwin Planetarium

Santa Barbara Museum
of Natural History
2559 Puesta del Sol Road
Santa Barbara, CA 93105-2936
Phone: (805) 682-4711
Fax: (805) 569-3170

With a 24' dome, seating for 60, and a Spitz A3P Star Projector, the Gladwin Planetarium at the Santa Barbara Museum of Natural History presents public shows on weekends, holidays, and Wednesday afternoons during the school year, and every day during major school breaks. The planetarium also presents shows to area school groups (call for details). Admission to the museum is $5 for adults,

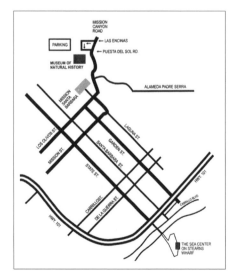

$4 for seniors and youths (12-23), $3 for children under 12. Admission to the planetarium is an additional $1.

The Gladwin Planetarium is part of the museum's Alice Tweed Tuohy Astronomy Center, which also features the E. L. Weigand SpaceLab. The SpaceLab has a collection of interactive astronomy exhibits that provide challenges for visitors of all ages and backgrounds. SpaceLab hours are from 9 a.m. to 5 p.m. weekdays, 10 a.m. to 5 p.m. on weekends and holidays.

The Museum of Natural History is also home to the Palmer Observatory (see separate entry).

Glenfield Planetarium

Glenfield Middle School
25 Maple Avenue
Montclair, NJ 07042-4513
Phone: (201) 509-4174
Fax: (201) 509-4179
Internet: www.interactive.net/~upper/planetar.html

Founded in 1984, the Glenfield Planetarium features a Minolta MS-8 star projector and hosts over 6000 visitors annually. A summer program is available for students, and several community events are scheduled throughout the year.

Goddard Space Flight Center

Visitor Center
Greenbelt, MD 20771
Code 130
Phone: (301) 286-8981
Group Tours: (301) 286-8103
TDD: (301) 286-8103
and:
Educator Resource Center
Code 130.3
Greenbelt, MD 20771
Phone: (301) 286-8570
TDD: (301) 286-8570

The Goddard Space Flight Center Visitor Center is open to the public daily from 9 a.m. to 4 p.m. (closed on Thanksgiving, Christmas, and New Year's Day). There is no admission charge, and parking is free. Visitors requiring special assistance are asked to contact a staff member at the Visitor Information desk or call (301) 286-8981. Hearing impaired visitors may schedule tours through a Telephone Device for the Deaf (TDD) at (301) 286-8103.

Public tours are offered Monday through Saturday at 11:30 a.m. and 2:30 p.m. Tours begin at the information desk and include stops at the NASA Communications Network and the control centers for the Hubble Space Telescope. Group tours for 15 or more people require reservations made a minimum of one month in advance. Bus group driving tours are available with advance notice. Tours last approximately 90 minutes.

The Goddard Space Flight Center offers an Educator Resource Center that serves educators in the northeastern states of Connecticut, Delaware, Maine, Maryland, Massachusetts, New Hampshire, New Jersey,

New York, Pennsylvania, Rhode Island, Vermont, and the District of Columbia.

Located at the Goddard Visitor Center, the Educator Resource Center is a place where educators can come and use NASA resources to develop their aerospace education programs. Here educators can research, gather ideas and duplicate audiovisual materials. Materials available at the resource center reflect NASA's research and technology development and relate to such curricula areas as astronomy, Earth science, aeronautics, mathematics, physical science and life science. Educators in disciplines other than science and mathematics are encouraged to explore ways in which aerospace materials may be incorporated into their classroom lessons.

Available resources include NASA publications, videotapes, 35mm slides, audio cassettes, filmstrips, teacher's guides with activities, reference books and literature, bibliographies, computer software, and Internet access. More than 800 videotapes are available for duplication at the Educator Resource Center. Programs are five to sixty minutes long and include many of the NASA films. Educators may use the centers video recorder to copy these programs, but must provide their own blank tapes (3/4 inch Umatic and ½ inch VHS format). Reservations are necessary to use the duplicating equipment.

Goldendale Observatory

1602 Observatory Drive
Goldendale, WA 98620
Phone: (509) 773-3141
E-mail: goldobs@gorge.net

The Goldendale Observatory State Park Interpretive Center features a 24-½" reflecting Cassegrain telescope, one of the nation's largest public telescopes. The telescope is available to visitors wanting to learn more about celestial bodies and to amateur astronomers working on special observing projects. Visitors can also bring their own telescopes or borrow a portable telescope during public visiting hours.

Tours are available for up to 49 people and can include slide shows, exhibits, films, and telescope viewing. Lectures on various astronomical phenomena are presented and the observatory often becomes a gathering point for comet watching, eclipse observation and other events. The facility also features a reference library, and an indoor the-

Photo: Goldendale Observatory State Park

ater/meeting room. The park is open Wednesday through Sunday, 2 - 5 p.m. and 8 p.m. - Midnight, from April 1 through September 30. From October 1 through March 31 visitors may drop in on Saturday and Sunday afternoons from 1 - 5 p.m. and Saturday evenings from 7 - 9 p.m. Afternoon

and evening visits are also available Wednesday through Friday and on Sunday evenings by appointment.

one of the nation's largest public telescopes

Griffith Observatory

2800 East Observatory Road
Los Angeles, CA 90027
Phone: (213) 664-1181 (office)
Hours & Show Times: (213) 664-1191
The Sky Report: (213) 664-8171
Friends of the Obs: (818) 846-3686
Fax: (213) 663-4323
E-mail: jmosley@earthlink.net
Internet: www.GriffithObs.org

The spectacular Griffith Observatory is visited by nearly 2 million people each year (over 50 million since open-

visited by nearly 2 million people each year

ing its doors in 1935), making it the seventh most popular tourist attraction in Southern California. The Observatory sits on the southern slope of Mount Hollywood and offers a stunning view of the Los Angeles basin. On the front lawn you'll see the Astronomers Monument, which honors Hipparchus, Copernicus, Galileo, Kepler, Newton, and Herschel, six of the greatest as-

Photo: Griffith Observatory

Photo: Griffith Observatory

tronomers of all time. Inside you'll have the opportunity to tour the *Hall of Science*. Exhibits include the Foucault Pendulum, the Solar Telescope, the 6-foot Earth and Moon Globes, the Meteorites exhibit, the Giant Tesla Coil (an electrifying experience), a Cosmic Ray Cloud Chamber, and so much more.

The centerpiece of the Observatory is its planetarium. Shows are presented at specific times each day, with school shows at 10:30 a.m. Tuesday through Friday during the school year (a school show brochure is available to teachers). The planetarium show is the only part of the Observatory that has an admission charge.

Griffith Observatory also offers a Bookshop and Space Stop gift area. You'll find an assortment of books, posters, astronomical slides, and gifts. The Observatory also has a mail order division where you can order a subscription to the *Griffith Observer*, the Observatory's monthly magazine; purchase an *Astrorama*, the Observatory's own planisphere which you can use to find your way around the heavens; choose from a selection of *Golden Guide Books*, or purchase a variety of other items.

An outstanding site.

Hansen Planetarium

15 South State Street
Salt Lake City, UT 84111-1590
Office: (801) 538-2104
Fax: (801) 531-4948
Star Show Schedule: (801) 538-2098
Laser Show Schedule: (801) 363-0559
Starline Info: (801) 532-STAR
Internet: www.utah.edu/
planetarium/

The Hansen Planetarium produces and presents educational programs, shows, and exhibits at the planetarium, as well as in local schools, and at various public gatherings.

Harvard-Smithsonian Center for Astrophysics

60 Garden Street
Cambridge, MA 02138
Phone: (617) 495-7461
Astronomy Information: (617) 496-STAR

For details, see listing in the *Organizations* section.

Hayden Planetarium

Museum of Science
Larry Schindler, Planetarium Director
Science Park
Boston, MA 02114-1099
Phone: (617) 723-2500
Fax: (617) 589-0454
E-mail: schindler@a1.mos.org
Internet: www.mos.org

The Charles Hayden Planetarium at the Museum of Science in Boston is more than just a planetarium. It is a multimedia, interactive astronomy-exploring facility. The planetarium is home to the world's first closed-captioning system for hearing-impaired visitors.

Each ergonomic seat has a three button key pad mounted on the armrest which allows audiences to interact with the programs. This Audience Response System is an integral component of planetarium shows and contributes to the dynamic, participatory feel of the planetarium experience. Red maplights are also mounted in the armrests, enabling visitors to read star charts or take notes without disrupting their dark-adapted vision.

The Zeiss Star Projector, a rotating star simulator, stands in the center of the 240-seat planetarium. The Zeiss projector simulates 9,000 stars and 29 constellations onto the 60' dome to create a stunningly realistic night sky. Computer-generated images of planets, galaxies, black holes and spacecraft

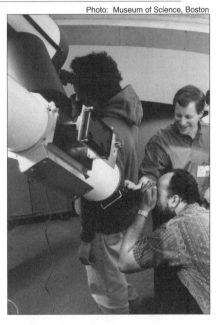

Photo: Museum of Science, Boston

The Museum's Gilliland Observatory

add a heightened level of dynamism to the cosmic scenes.

The Planetarium's multi-media system features 108 special effect projectors, five large-format video projectors, 72 slide projectors and three laser disc players, enabling the planetarium staff to produce a countless number of visual images.

Located on the Museum's Parking Garage roof is the Gilliland Observatory which features two telescopes: a 7" refractor and a 12" Schmidt-Cassegrain. During Friday night viewing hours, these instruments provide visitors with spectacular views of the universe, including sunspots, stars, planets, lunar craters, asteroids, comets and distant galaxies. The observatory can also transmit "live" pictures of space objects directly from the telescopes to the planetarium, where visitors can enjoy cosmic images from the comfort of the theater. The planetarium offers star shows, laser shows set to the tunes of rock music, and seasonal, multimedia programs that delve into the latest cosmic topics.

Showtime and ticket information for the Museum of Science and the Charles Hayden Planetarium are available by calling (617) 723-2500. School group information is available by calling (617) 723-2511. The Museum of Science and the Charles Hayden Planetarium are open seven days a week. Planetarium and Laser shows tickets are $7.50 for adults, $5.50 for children ages 3-14 and seniors 65 and up. The Museum of Science and the Planetarium are wheelchair accessible. The planetarium offers assistive listening devices and a captioning system for hearing-impaired visitors.

Hayfield Secondary School Planetarium

Tony Klepic, Planetarium Teacher
7630 Telegraph Road
Alexandria, VA 22310-1180
Phone: (703) 992-7537

The Hayfield Secondary School planetarium opened in fall of 1969 and features a Spitz A4 projector, a 30-foot diameter dome, and 75 seats in a chevron pattern.

Herndon High School Planetarium

Mary Blessing, Planetarium Teacher
700 Bennett Street
Herndon, VA 20170-3270
Phone: (703) 810-2335

The Herndon High School planetarium opened in the fall of 1967 and features a Spitz A3P projection system, 30-foot diameter dome, and permanent seating for 48 (plus 30 folding chairs).

Holcomb Observatory & Planetarium

Butler University
4600 Sunset Avenue
Indianapolis, IN 46208
Information: (317) 940-9333
Tours: (317) 940-9352
Internet: tekel.butler.edu/holcomb/index.html

Holcomb Observatory and Planetarium, located at the south edge of Holcomb Gardens at Butler University, is open to the public for weekend tours on Friday and Saturday evenings throughout the academic year. Presentations are at 7 p.m. and 8:15 p.m. Private tours are available for groups of 15 or more people by calling (317) 940-9352 for reservations. The facility offers both planetarium shows and the opportunity to view celestial objects through the largest telescope in Indiana—a 38-inch Cassegrain Reflector telescope.

Holt Planetarium

University of California
Lawrence Hall of Science #5200
Berkeley, CA 94720-5200
Phone: (510) 643-5082
Fax: (510) 642-1055
E-mail: agould@uclink4.berkeley.edu
Internet: www.lhs.berkeley.edu

The William K. Holt Planetarium is located in the Lawrence Hall of Science on the University of California, Berkeley campus. Holt Planetarium has developed an international reputation as a leader in design of effective activity-oriented planetarium programs—programs in which audience members actively participate. The 6.8 meter dome features comfortable cushioned bench-type seats accommodating 27 visitors. School shows are presented every weekday during the school year and three public shows are presented every Saturday and Sunday, holidays, and daily during the summer. Planetarium tickets are $2.00 each. Programs are recommended for ages 8 and up. Children under the age of 6 can be accommodated in special reserved workshops.

The Holt Planetarium has published many of its unique audience-participatory shows in the series Planetarium Activities for Student Success (PASS), authored and edited by Cary Sneider, Alan Friedman and Alan Gould. Designed for both experienced planetarium instructors and teachers using a planetarium for the first time, this series provides a wealth of ready-to-go ideas and practical suggestions for planning and presenting entertaining and educationally effective programs for people of all ages. Eight of the volumes each contain a complete planetarium program and related classroom activities. For more information about PASS, call 510-642-5863. The volume titles are: (1) Planetarium Educator's Workshop Guide, (2) Planetarium Activities for Schools, (3) Resources for Teaching Astronomy & Space Science, (4) A Manual for Using Portable Planetariums, (5) Constellations Tonight, (6) Red Planet Mars, (7) Moons of the Solar System, (8) Colors From Space, (9) How Big Is the Universe? (10) Who "Discovered" America? (11) Astronomy of the Americas (12) Stonehenge. To order, call Eureka! at 510-642-1016.

Hook Astronomical Observatory

Dr. Robert C. Howe, Director
Department of Geography, Geology, and Anthropology
Indiana State University
Terre Haute, IN 47809
Phone: (812) 237-2271
Fax: (812) 237-8029
E-mail: gehowe@scifac.indstate.edu
Internet: baby.indstate.edu/gga

Operated by the Department of Geography, Geology, and Anthropology, the Hook Astronomical Observatory is located on top of the Science Building near the center of downtown Terre Haute. The observatory is equipped with a 10" Meade LX 200, several other telescopes and a digital camera system for astrophotography. The observatory is open to the public on Tuesday and Friday nights when it is clear (there is no charge for admission).

Hummel Planetarium

Jack K. Fletcher, Director
Eastern Kentucky University
Richmond, KY 40475
Phone: (606) 622-1547
Fax: (606) 622-6205
E-mail: hplfletc@acs.eku.edu.
Internet: eagle.eku.edu/
campus_tour/campus_tour.html.

The Armin D. Hummel Planetarium features a 20.6-meter dome, seating for 164 and a gift shop. Located on the campus of Eastern Kentucky University, the planetarium serves the campus as well as all public and private, elementary and secondary schools and the general public. School programs are presented Monday through Friday by reservations only. Admission is $2.50 per student. Approximately 35,000 elementary and secondary school students visit the planetarium each year. Programs open to the general public are presented every Thursday and Friday at 7:30 p.m. and every Saturday and Sunday at 3:30 p.m. and 7:30 p.m. Admission to public programs is $3.50 for adults, $3.00 for senior citizens and students, and $2.75 for children 12 and under. The planetarium office is open Monday through Friday from 8 a.m. to 4:30 p.m.

Indiana University of Pennsylvania Planetarium

Connie J. Sutton, Director
Indiana University of PA
Geoscience Department
114 Walsh Hall
Indiana, PA 15705
Phone: (412) 357-2379
Fax: (412) 357-5700
E-mail: cjsutton@grove.iup.edu

The Indiana University of Pennsylvania (IUP) Planetarium features a Spitz A3P housed in a forty-foot dome with 75 uni-directional seats. It is used by IUP students enrolled in astronomy courses and is available to outside organizations such as school groups, scout troops, and civic organizations. A $25 fee is assessed for a program. Programs are geared to the age level and knowledge level of the audience. Small telescopes are also available for observation use with the groups depending on weather and time of day/ night.

Jesse Besser Museum Planetarium

Tom Gougeon, Coordinator
491 Johnson Street
Alpena, MI 49707
Phone: (517) 356-2202
Fax: (517) 356-3133
E-mail: jbmuseum@northland.lib.mi.us
Internet: www.oweb.com/upnorth/
museum/home.html

The Jesse Besser Museum Planetarium features a Spitz A3P planetarium instrument under a 30 foot dome with seating for 60. Programming is offered for elementary, secondary and college level students as well as for the general public. Reservations are required for all school and other special groups. Public programs are offered on Sunday afternoons at 1:00 and 3:00 p.m. During the months of July and August, additional public programs, featuring the summer sky, are presented on Thursday evenings at 7:30 p.m. Admission to the planetarium is 75 cents in addition to Museum admission, which is $2.00 for adults, $1.00 for senior citizens and students under 18, and $5.00 for a family.

Jet Propulsion Laboratory

4800 Oak Grove Drive
Pasadena, California 91109
Phone: (818)354-4321
Public Affairs: (818) 354-5011
Internet: www.jpl.nasa.gov

Jet Propulsion Laboratory (JPL) is the lead U.S. center for robotic exploration of the solar system. JPL spacecraft have visited all known planets except Pluto (a Pluto mission is currently under study for the late 1990s).

JPL's main site occupies 177-acres at the foot of the San Gabriel Mountains near Pasadena, California, 19 kilometers (12 miles) northeast of Los Angeles. JPL manages the worldwide Deep Space Network, which communicates with spacecraft and conducts scientific investigations from its complexes in California's Mojave Desert near Goldstone; near Madrid, Spain; and near Canberra, Australia.

Johnson Planetarium

James Beaber, Director
Jefferson County School District
200 Kipling Street
Lakewood, CO 80226
Phone: (303) 237-4786
Fax: (303) 232-3827
E-mail: jbeaber@teal.csn.net

The Robert H. Johnson Planetarium was built in 1962 through a National Science Foundation grant as part of the systemic effort to infuse more science education into the public schools. The facility features a 10.1 meter dome, concentric seating for 125, 70 special effects projectors, and four video projectors. The original star projector was a Goto MK I. In 1995, the school district replaced the original projector with a new Zeiss ZKP 3 star projector. This projector, purchased through Seiler Instruments, provides an accurate and beautiful sky, with stars to magnitude 6.3. The accuracy of the machine is necessary to complement the often clear and beautiful Colorado skies the students observe.

The educational mission for the planetarium is to provide astronomy instruction for the students of the Jefferson County School District, which is the largest district in the state of Colorado. Planetarium programs are included in the district science curriculum for grades 3-6. The planetarium has approximately 47,000 student visits per year. The fee is $2.00 per student per visit. Out-of-district schools are allowed to attend the planetarium as the schedule permits. The fee is the same.

In addition to the planetarium facility, the director oversees two mountain observatories located in the school district's two outdoor education lab schools. The Windy Peak Outdoor Lab School features an astronomy classroom with a concrete apron for telescope viewing, two Celestron C 14 telescopes, two C 8's, and a richest field telescope. The Mt. Evans Outdoor Lab School features an observatory/classroom and a 22" Celestron telescope that is owned and jointly used by the Denver Museum of Natural History's Gates Planetarium. Two C 8's, a five-inch refractor, and a C 5 complement the 22" telescope. Each 6th-grade class in the district spends a week at one of the lab schools as part of the science curriculum. The observatories are used primarily for observations by the students while at lab school.

Johnson Space Center

2101 NASA Road 1
Houston, TX, 77058
Phone: (281) 483-0123
Information: (281) 483-8693
Community Affairs: (281) 244-8024

Astronaut speakers may be requested through the Astronaut Appearances Office

The Lyndon B. Johnson Space Center (JSC) was established in September 1961 as NASA's primary center for design, development, and testing of spacecraft; selection and training of astronauts; and planning and conducting manned space flight missions.

Johnson Space Center is adjacent to Clear Lake, about 20 miles southeast of downtown Houston via Interstate 45. A number of services are available through JSC, including astronaut speakers, exhibit loans, and a Teacher Resource Center.

Astronaut speakers may be requested through the Astronaut Appearances Office at 281-244-8866. General NASA speakers (scientists, engineers, etc.) may be requested through Community Affairs at 281-244-8024. To find out about borrowing JSC exhibits, call the Exhibits Manager at 281-483-8622. Tours can be arranged at Space Center Houston, JSC's official visitor center (see the Space Center Houston listing in the *Places of Interest* section). Space Center Houston is located adjacent to JSC and is managed in partnership with non-profit private sector organizations. NASA souvenirs may be purchased from the Space Trader Gift Shop at 1-800-746-7724. Individual astronaut lithographs may be obtained by writing to Mail Code CB, Johnson Space Center, 2101 NASA Road 1, Houston, TX 77058-3696.

The Johnson Space Center Teacher Resource Center (TRC) serves educators in an eight-state region (Texas, Oklahoma, Colorado, Kansas, Nebraska, New Mexico, North Dakota and South Dakota). The phone number for this TRC is 281-486-8696.

Jordan Planetarium and Observatory

Alan Davenport, Director
5781 Wingate Hall
University of Maine
Orono, ME 04469-5781
Phone: (207) 581-1341
E-mail: AlanD@maine.maine.edu
Internet: www.ume.maine.edu/
~lookup/

The Maynard F. Jordan Planetarium is located on the second floor of Wingate Hall at the University of

Photo: Jordan Observatory

Maine and features a 20-foot dome, a Spitz model 373, multi-media equipment, sound systems, slide projectors, special effect projectors, and seating for 45. Programs explore the solar system, distant galaxies, and the constellations. Private showings can be arranged for school field trips, private parties, and clubs. The planetarium store sells astronomy-related toys, souvenirs, and other materials. For teachers, the planetarium has a large selection of astronomical resource materials.

The Maynard F. Jordan Observatory was installed in 1901 and is used by students for their course work in astronomy. The observatory houses an Alvan Clark refractor telescope and is available to the general public during special hours.

Kansas Cosmosphere and Space Center

1100 North Plum
Hutchinson, KS 67501-1499
Phone: (316) 662-2305
Fax: (316) 662-3693
Internet: www.cosmo.org

Housing the largest space artifact collection outside the Smithsonian Institution's Air and Space Museum

With a humble beginning as the Hutchinson Planetarium back in 1962, the Kansas Cosmosphere and Space Center has evolved into a journey of discovery about man's reach for the stars. Housing the largest space artifact collection outside the Smithsonian Institution's Air and Space Museum, the Cosmosphere's $250 million artifact collection is used to tell the comprehensive story of the space race between the United States and the former Soviet Union, from early rocketry to the present and future of space exploration. A combination of exhibits, IMAX Dome films and other educational programs offers insight into today's American and Russian cooperative exploration of space.

The crown jewel of the collection is the actual Apollo 13 command module "Odyssey," which is currently being restored to its original condition by Space Works, Inc., a space artifact restoration and replication facility and wholly-owned subsidiary of the Cosmosphere, also located in Hutchinson, Kansas.

Other major exhibits featured in the Cosmosphere's Hall of Space Museum include the most significant collection of Russian space artifacts outside Russia and rare and authentic German V1 and V2 rockets. The country's largest collection of American and Russian space suits, as well as Mercury, Gemini and Apollo spacecraft, are also on display or being readied for display.

Visitors to this impressive 105,000 square foot facility are welcomed by a 30-ton SR-71 Blackbird, one of only a dozen such aircraft on display in the world. The Mach 3+ spy plane is permanently mounted in the Cosmosphere's lobby along with a Northrop T-38 Talon and a full-scale replica of the Space Shuttle.

The Apollo 13 Command Module, now being restored at the Kansas Cosmosphere

Kelce Planetarium

Department of Physics
Pittsburg State University
1701 S. Broadway
Pittsburg, KS 66762
Phone: (316) 235-4391
Fax: (316) 235-4050
E-mail: tvangord@mail.pittstate.edu

Located at the corner of Joplin and Cleveland Streets, the L. Russell Kelce Planetarium offers weekly programs to the general public covering a variety of topics. Programs for area elementary and secondary schools and universities are also available.

Kelly Space Voyager Planetarium

Doug Baldwin, Director
Discovery Place
301 North Tryon Street
Charlotte, NC 28202
Phone: (704) 372-6261 ext. 451
Fax: (704) 337-2670
E-mail: web6251@charweb.org
Internet: www.discoveryplace.org/

The Kelly Space Voyager Planetarium is a part of Discovery Place, a hands-on children's science museum. The planetarium is combined with an OMNIMAX theater to make the dome five stories high and 79 feet in diameter. It was listed in the 1995 Guinness

Book of World Records as the largest planetarium dome in the United States. At the heart of the Kelly Planetarium is the Spitz Voyager Starball which accurately projects 10,164 stars with their correct size, color and brightness. In addition, 40 slide projectors create beautiful full screen panoramas. The planetarium is open to the public, and school groups often take advantage of the theater. Programs are presented seven days a week. Admission to the planetarium is $2.00 with admission to Discovery Place. If you only wish to visit the planetarium, the cost is $6.50 (ages 13-59), $5.00 for youths (ages 6-12) and seniors (ages 60+), $2.75 for children 3-5, and free for children under the age of 3.

Kennedy Space Center

Visitor Complex
Kennedy Space Center, FL 32899
NASA: (407) 867-4636
Visitor Complex: 1-800-572-4636
Internet: www.kscvisitor.com

The Visitor Complex at John F. Kennedy Space Center features a modern complex of exhibit halls, theaters, and supporting amenities, such as indoor and outdoor exhibits and displays featuring the spacecraft, rockets, and missions of the U.S. space program. IMAX movies offer a spectacular view of space as seen by the astronauts. With such a lineup, it's not surprising that the Kennedy Space Center attracts close to 2.5 million visitors a year.

The Visitor Complex offers the only back-to-back twin IMAX theater complex in the world, showcasing three exciting, large-format motion pictures on a screen over 5 stories tall. *Mission to Mir* is a 40-minute, giant-screen tour of Russia's space station that gives viewers a unique look inside the weightless home in space that has been occupied by international teams of scientists since 1986. *The Dream is Alive* is a spectacular space adventure and one of the more popular IMAX presentations at the Visitor Complex. The huge screen and thunderous sound system make you feel like you're standing on the launch pad during a space shuttle launch! *L5, First City in Space*, is a new 3-D space movie and the first IMAX film to accurately depict a future space settlement, uses real NASA footage and data. Experience this fictional orbiting community's perilous interplanetary quest for water and take a thrilling flight through the mountains and valleys on the surface of Mars, all in incredible 3-D.

Photo: NASA

Photo: NASA

Those wishing to visit the Space Shuttle launch site may board a bus at the Visitor Complex and take the Red Tour that highlights Space Shuttle Launch Pads A and B, the massive Vehicle Assembly Building, and an authentic 365 foot long Saturn V moon rocket housed in the new Apollo/Saturn V Center. Also available is the two hour Blue Tour of Cape Canaveral Air Station, where the U.S. space program began in the 1950's and 1960's.

Additional activities at the Visitor Complex include an outdoor Rocket Garden where you can view a collection of eight authentic rockets from past eras; a walk-through exhibit *Satellites and You* that offers a 45 minute journey through a simulated future space station; and *The Gallery of Spaceflight* which dedicates an entire building toward space exploration, displaying authentic Mercury and Gemini capsules. The 250 piece NASA Art Gallery in Galaxy Center highlights a complete range of artistic styles.

Kennedy Space Center also offers the Center for Space Education which provides extensive facilities to aid teachers, such as a large number of free aerospace publications, plus videotapes, 35mm slides, and text data that can be copied on site. The Exploration Station, located in the Center for Space Education, provides educational programs and hands-on activities that explain the principles of rocketry and space science. Students have the opportunity to work with the actual hardware used for space missions.

Souvenir shoppers will want to take some time to look over The Space Shop's 8,000+ space-related items ranging from clothing and collectibles to photos and videos. It's the perfect place to wrap up your visit. For those who can't make the trip in person, The Space Shop is available online!

The KSC Visitor Complex is located on 70 acres of Florida's Space Coast, one hour east of Orlando (two miles south of Titusville, Florida, off U.S. Highway 1). There are no admission or parking fees to enter the Visitor Complex and special assistance is available for the hearing-impaired (contact TDD #: (407) 454-4198 in advance) and physically disabled visitors. Free wheelchairs are provided and all exhibits and tours are handicapped-accessible. Taking one step beyond superior customer service, the Visitor Complex also offers free camera and stroller use, and a free pet kennel service. The Visitor Complex is open every day of the year except Christmas.

Space Shuttle Launches

The Visitor Complex at Kennedy Space Center sells approximately 1,500 tickets to board buses that transport visitors to a viewing site to watch the Space Shuttle launch. Prices are $10.00 for a Launch Viewing Opportunity (LVO), which includes an LVO ticket, and an IMAX movie presentation. Tickets may be purchased 5 days prior to launch and must be picked up in person prior to launch day.

Outside the Kennedy Space Center, prime viewing seats are inland along U.S. Highway 1 in the city of Titusville and along Highway A1A in the cities of Cape Canaveral and Cocoa Beach on the Atlantic Ocean. A limited number of Launch Viewing car passes are issued by NASA by writing several months in advance to: NASA VISITOR SERVICES, MAIL CODE: PA-PASS, KENNEDY SPACE CENTER, FL 32899, or by calling 407-867-6000 for more information. Updated launch information is available by calling NASA at (407) 867-4636 or the Visitor Complex at 1-800-572-4636.

Kennon Observatory

Department of Physics & Astronomy
University of Mississippi-Oxford
University, MS 38677
Phone: (601) 232-7046
E-mail:
physics@beauty1.phy.olemiss.edu
Internet: beauty1.phy.olemiss.edu/
homepage.html
Internet: www.olemiss.edu/depts/
physics_and_astronomy/

The Kennon Observatory includes the following equipment: 15" f/12 Refractor with German Equatorial Mount (Grubb 1893); 7" f/15 Maksutov Reflector with Fork Mount (Questar ~1988); 8" f/6 LX200 Schmidt-Cassegrain with Fork Mount (Meade 1996); 12" f/10 LX200 Schmidt-Cassegrain with Fork Mount (Meade 1996); 25" f/5 Newtonian Reflector with 2-axis Poncet Drive (Obsession 1997); and SBIG Model ST7 765x510 pixel CCD Camera (SBIG 1996).

Kirkpatrick Planetarium

Wayne Wyrick, Director
Omniplex Science Museum
2100 NE 52nd Street
Oklahoma City, OK 73111
Phone: (405) 424-5545
Fax: (405) 424-5106
E-mail: wizardwayne@juno.com
Internet: www.ionet.net/~omniplex

The observatory on the roof houses the Maguire Telescope, the largest public access telescope in the state.

The Kirkpatrick Planetarium offers daily programs under a 40-foot dome with seating for 120. The planetarium is located in a museum complex known as "The Kirkpatrick Air Space and Science Museum at Omniplex," which features numerous attractions. In addition to the planetarium, visitors can visit the Air Space Museum and "orbit" in a Mercury simulator or prowl the skies in an F-16 flight simulator. Omniplex Science Museum offers more that 300 hands-on science exhibits and activities, including a fully functional television studio. Behind the main building, visitors may walk through three acres of gardens and greenhouses. Numerous art, cultural and historical galleries round out the complex including such attractions as the International Photography Hall of Fame, Red Earth Indian Center, and the Ntu African gallery. The museum complex is open daily except for Christmas and Thanksgiving. Some one hundred different programs are offered for school children on field trips covering a great variety of subject matters. The observatory on the roof houses the Maguire Telescope, the largest public access telescope in the state.

The Kirkpatrick Planetarium is the host for the Oklahoma City Astronomy Club (see listing under *Organizations*). Its one hundred and three members run Canyon Camp Observatory forty miles west of town under dark skies. The Astronomy Club meets on the second Friday of each month at 7:00 p.m.

more that 300 hands-on science exhibits and activities

Kirkwood Observatory

Indiana University
Swain West 319
Bloomington, IN 47405
Phone: (812) 855-6911
E-mail: request@astro.indiana.edu
World Wide Web:
www.astro.indiana/edu

Kirkwood Observatory features a 12-inch refractor telescope and a solar telescope. The observatory offers public viewing opportunities every Wednesday evening, March through October, when classes are in session (weather permitting). No reservations are necessary. Group showings are available by contacting the Astronomy Department at (812) 855-6911 or via e-mail.

The observatory offers public viewing opportunities every Wednesday evening, March through October

Koch Science Center and Planetarium

Evansville Museum of Arts & Science
411 S.E. Riverside Drive
Evansville, IN 47713
Phone: (812) 425-2406
Recording: (812) 421-7510

The Koch Planetarium at the Evansville Museum of Arts and Science has a year round schedule of shows. Afternoon programs are offered twice daily on weekends and daily during the summer. Solar observing using the museum's Solaris H-Alpha Telescope is offered on many days. The museum also has a gift shop which stocks astronomical books, unusual gifts, and other science related items. The museum is open Tuesday through Saturday from 10 a.m. to 5 p.m. and on Sunday from noon to 5 p.m. The museum is closed on Mondays and holidays.

Photo: Koch Science Center and Planetarium

Lakeview Museum of Arts & Sciences

Sheldon Schafer, Director of Science
Programs & Facilities
1125 West Lake Ave
Peoria, IL 61614
Phone: (309) 686-7000
Fax: (309) 686-0280
E-mail: sls@bradley.bradley.edu
Internet: www.lakeview-museum.org

Home of the world's largest model of the Solar System. For a cyberspace tour, visit web site www.bradley.edu/las/phy/solar_system.html.

Link Planetarium

Roberson Museum and
Science Center
30 Front Street
Binghamton, NY 13905-4779
Phone: (607) 772-0660
Fax: (607) 771-8905
E-mail: linkpl@juno.com

The Roberson Museum and Science Center houses the Link Planetarium and offers programs for the general public on Friday nights, weekends, and school holidays. The planetarium has seating for 55 under a 30-foot dome and uses a Spitz Star Projector. Weekly star shows examine the current night sky and provide images from recent discoveries and space exploration. The museum also features the Kopernik Space Education Center, a scientific laboratory for hands-on training of teachers, students, and the general public. For those interested in joining an astronomy club, The Kopernik Astronomical Society is a constituent organization of the Roberson Museum and Science Center. Museum admission is $4 for adults, $3 for students. Planetarium admission is $1 additional. Call for show dates and times.

Longway Planetarium

Flint Cultural Center
1310 East Kearsley Street
Flint, MI 48503
Star Show Info: (810) 760-1181
Laser Light Show Info: (810) 760-7511
24-Hour Night Sky Info: (810) 760-5462
E-mail: mgardner@flint.org
Internet: www.flint.org/longway

The Robert T. Longway Planetarium is one of the top planetarium theaters in the United States and features 285 seats under a 60-foot dome. A Spitz Model B Projector is the heart of the planetarium and can accurately reproduce the sky with over 4,000 stars. Hundreds of auxiliary projectors help recreate the sun, moon, planets, comets, and other celestial phenomena.

Public programs at Longway Planetarium are 45 to 50 minutes long and cover subjects such as the Solar System, Galaxies, and man's exploration of space. Laser programs are also presented throughout the year, and special programs are available to the general public, school groups, and other interested organizations. For those interested in supporting the work of this fine facility, membership in *The Friends of the Robert T. Longway Planetarium* is available.

Longwood Regional Planetarium

Joseph E. Caprioglio, Director
Longwood Middle School
41 Yaphank Middle Island Road
Middle Island, NY 11953
Phone: (516) 345-2741
Fax: (516) 345-9296
E-mail: gmastrion@worldnet.att

The Longwood Regional Planetarium features a 30-foot dome, seating for 63, and a nice selection of programs for grades K-12. Public programs are occasionally offered during the evening and on weekends. Evening shows are followed by celestial observing sessions, weather permitting.

Lowell Observatory

1400 West Mars Hill Road
Flagstaff, AZ 86001-4499
Phone: (520) 774-2096
Internet: www.lowell.edu

Lowell Observatory, founded in 1894, is not only a thriving center for pioneering research and education in astronomy, but is a historical landmark and museum. Many discoveries have been made at Lowell Observatory, the most famous of these being the 1930 discovery of the ninth planet, Pluto. Currently, Lowell Observatory astronomers are studying everything from the moon Io, which circles Jupiter, to binary stars. The observatory also boasts the world's leading expert in asteroids, focusing on asteroids that might collide with the Earth. Lowell Observatory has also teamed up with the U.S. Navy to build the most advanced optical interferometer in the world. When completed, the optical interferometer will be able to measure star positions 100 times better than has previously been done from the Earth's surface.

Lowell Observatory offers programs for the entire family. Admission to the Steele Visitor Center includes a tour of the entire facility featuring the telescope with which Pluto was discovered. The visitor's center consists of a multimedia lecture hall, a gift shop and a diverse series of exhibits entitled "Tools of the Astronomer." The exhibits offer interactive displays that demonstrate current astronomical technology. Displays include an unprecedented "simulated observatory," a full-size research telescope, various hands-on demonstrations and exhibits encouraging the pursuit of astronomy as a hobby or career.

Marshall Space Flight Center

Huntsville, AL

The Marshall Center occupies 1,800 acres in Huntsville, Alabama. Its primary mission is to develop and maintain space transportation and propulsion systems and conduct microgravity research. In addition to being NASA's lead center for transportation systems and microgravity research, Marshall hosts NASA's Centers of Excellence in science research in materials, biotechnology, global hydrology and climate studies, astrophysics, and payload utilization. In addition, Marshall houses Centers of Excellence in propulsion, avionics, space environment and optics technologies.

McDonnell Planetarium

St. Louis Science Center
5050 Oakland Avenue
St. Louis, MO 63110
Phone: (314) 289-4400
Internet: www.slsc.org

Located in the Forest Park Building of the St. Louis Science Center, the McDonnell Planetarium features family and night sky shows, plus weekend laser light shows. State-of-the-art technology includes 3-D images, six-channel surround-sound and a

Digistar® projector that creates moveable star fields, all under a 60-foot domed screen with seating for 212.

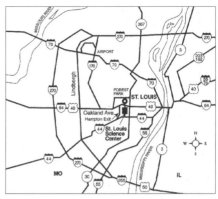

The Science Center also features two space galleries, *Space Xploration* and *The Strange Place of Outer Space.* Here you can transport to another planet, discover what makes black holes, quasars and pulsars, or explore space and time through popular television and films of the past 90 years. The galleries occupy 10,000 square feet on the upper and lower level of the Forest Park Building. The galleries feature hundreds of artifacts, including an actual Gemini space capsule.

Of course, the planetarium and space exhibits represent only a fraction of the Science Center's displays that help draw over 1.5 million visitors annually. In all, the St. Louis Science Center occupies 232,000-square-feet in three buildings connected by a bridge and a tunnel. The main building offers more than 500 permanent exhibits, a Discovery Room, educational facilities, and an OMNIMAX® Theater. Tickets are required for the OMNIMAX®, planetarium, laser light shows, Discovery Room and traveling exhibitions. For help in planning a class visit or for supplemental curriculum material, call (314) 533-8493. Private parties are available and can be arranged by calling (314) 533-8179.

The Science Center also features two space galleries where you can transport to another planet, discover what makes black holes, quasars and pulsars, or explore space and time.

Photo: St. Louis Science Center

McKeldin Memorial Planetarium

St. John's College
Annapolis, MD 21404
Phone: (410) 263-2371
E-mail: beall@sjca.edu

McKeldin Memorial Planetarium is located on the St. John's College Campus in Annapolis, MD. The projector is one of the original Spitz models. The planetarium is used mostly for instructional purposes by St. John's College faculty, but public programs are given on an occasional basis.

Merrillville Community Planetarium

Clifford Pierce Middle School, 2nd Floor
199 East 70th Avenue
Merrillville, IN 46410
Phone: (219) 736-4856
Internet: www.calunet.com/org/
MCP/index.html

The Merrillville Community Planetarium is part of the Merrillville Community School Corporation and features a 30-foot perforated aluminum dome, a Spitz Model 512 star projector, approximately one hundred projectors, and seating for 64.

Public programs are presented on Friday and Saturday evenings at 7:30 p.m. Admission to the public program is $2.00 for adults and $1.00 for children. Because of limited seating, reservations are recommended.

Michigan Space & Science Center

2111 Emmons Road
Jackson, MI 49203
Phone: (517) 787-4425
Fax: (517) 796-8632
E-mail: stewart_bailey@jackson.cc.mi.us

The Michigan Space & Science Center is a unique facility housing over $30 million in space artifacts from all of America's space programs, from the first Mercury-Redstone launches to the latest flights of the Space Shuttle. Housed under a 12,000 square foot geodesic dome, the facility uses both static and interactive exhibits to tell the story of the human exploration of the final frontier. Among the exhibits are the Apollo 9 Command Module, a Moon Rock brought back by Apollo 15, a Mercury-Redstone rocket, a Mercury-Atlas Control Console, Mercury, Gemini and Apollo space suits, numerous items from the Skylab program and a memorial to the crew of the Shuttle Challenger. The Center also displays many personal items donated by Michigan's Astronauts, including James McDivitt, Al Worden, Jack Lousma, Roger Chaffee and David Leestma. The Center also features a gift shop which is well stocked with hundreds of space and space-flight related items.

The Michigan Space & Science Center is visited by over 20,000 people a year, a majority of whom are elementary and secondary school students. In addition to school group visits, the Center also offers overnight camps on weekends for scout and youth groups and week-long summer day camps for students in the 2nd-7th grades.

The Center is open Tuesday through Saturday, 10 a.m. to 5 p.m. year-round and also Sunday 12 p.m. to 5 p.m. from May 1 through Labor Day. Admission is $4.00 for adults, $2.75 for students and senior citizens, and children 5 and under are Free. Discounts are available for groups of 15 or more with an advanced registration.

Minneapolis Planetarium

300 Nicollet Mall
Minneapolis, MN 55401
Phone: (612) 630-6150
Reservations: (612) 630-6155

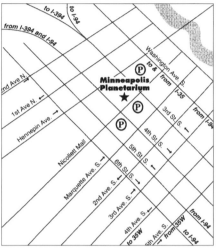

Located inside the downtown Minneapolis Public Library building, the Minneapolis Planetarium offers school programs Monday through Friday and public star shows on weekends and Thursday nights. The minimum size for school group reservations is 30. If you have fewer than 30, you may join a larger group. The star theater capacity ranges from 150 to 175. Admission is $4 for adults and teens, $2.50 for children 12 and under, and $2.25 each for groups of 10 or more.

The gift shop sells astronomy books, star charts, posters, t-shirts, home planetarium kits, and more.

Minolta Planetarium

De Anza College
21250 Stevens Creek Blvd.
Cupertino, CA 95014
Phone: (408) 864-8814
Fax: (408) 864-5643
E-mail: vonahnen@admin.fhda.edu
Internet: planetarium.fhda.edu

Over 20,000 people visit the Minolta Planetarium each year, taking advantage of programs that range from Family Astronomy Nights to Laser Light Shows. The planetarium offers school programs for preschool and kindergarten through 12th grade, including a Space Science Camp during the summer. The amphitheater seats 170 visitors beneath a 50-foot dome and dazzles audiences with state-of-the-art sound and star-projection equipment. Exhibits include a ro-

programs for preschool and kindergarten through 12th grade, including a Space Science Camp during the summer

tating earth display, model rocket display, and an ozone study exhibit, among others. Coming attractions will include a space exploration program, live NASA space transmissions, guest lectures, and other programs for the public. The Minolta Planetarium will also host a birthday party, anniversary, wedding, or corporate event under the 50-foot dome, which sounds like a lot of fun.

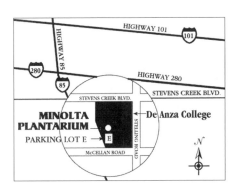

Montgomery College Planetarium

Takoma Avenue and Fenton Street
Takoma Park, MD 20912-4197
Phone: (301) 650-1463
Fax: (301) 650-1550 (college library)
E-mail: haroldw@umd5.umd.edu
Internet: myhouse.com/mc/
planet.htm

the planetarium sponsors a summer workshop for teachers

The Montgomery College Planetarium shows 1,834 naked-eye stars, the Milky Way, and the five naked-eye planets (Mercury, Venus, Mars, Jupiter, and Saturn) under a 24-foot dome with forty-two comfortable chairs. The planetarium is located on Fenton Street on the Takoma Park campus of Montgomery College. It is attached to the Science South building on the ground level and has a conspicuous silver-colored domed roof. An astrophysicist will answer questions about the universe. There is no admission charge for these public planetarium programs.

In addition to regularly scheduled shows, the planetarium sponsors a summer workshop for teachers, an astronomy course, grade-specific field trips for school groups, special shows for other groups, and monthly music/laser light shows.

The astronomy course offered by the planetarium allows participants to explore how stars are formed, know what a black hole is and where one might be located, find out what a neutron star is, find out what the sun is composed of, find out how the elements were formed, find out how and when the earth was formed, or learn the names of some of the constellations and the names of the brightest stars in them. The course, *Introduction to Astronomy at Montgomery College (T AS101)*, is the only course in the Washington Metro area that takes place within a planetarium. Senior citizens can register for this course for less than the cost of the text book on a space available basis one week before the class starts. *Introduction to Astronomy* is offered in the planetarium on Tuesdays and Thursdays from 1 p.m. to 3:45 p.m. Besides observing the stars in the planetarium, the class goes outside and uses telescopes on clear nights. This course is also available as a distant learning telecourse.

Morrison Planetarium

California Academy of Sciences
Golden Gate Park
San Francisco, CA 94118
Phone: (415) 750-7127 (office)
Recording: (415) 750-7141
Fax: (415) 750-7129
E-mail: planetarium@calacademy.org
Internet: www.calacademy.org/
planetarium

Located inside the California Academy of Sciences, the Morrison Planetarium, which opened to the public on November 8, 1952, features a 65' dome and concentric seating for 309. The planetarium is wheelchair accessible and has the Phonic Ear system for the hearing impaired. Museum admission is $8.50 for adults, $5.50 for youths (12-17) and seniors, $2 for children (4-11), and free of charge for children under 4. Planetarium admission is an additional $2.50 for adults, $1.25 for students (6-17) and seniors over 64. The planetarium has shows daily at 2 p.m. and on the weekends at 11 a.m., Noon, and at 1, 2, 3, and 4 p.m. Call for show titles.

Mueller Planetarium and Observatory

Cleveland Museum of Natural History
Coordinators:
Joseph DeRocher, Planetarium
Clyde Simpson, Observatory
1 Wade Oval Drive, University Circle
Cleveland, OH 44106-1767
Phone: (216) 231-4600
Fax: (216) 231-5919
(216) 231-9960
E-mail: jderoche@cmnh.org
         csimpson@cmnh.org
Internet: www.cmnh.org

The Ralph Mueller Planetarium contains a 10-meter dome with seating for 85. The heart of the planetarium is a Spitz A3P, installed in 1966. The Ralph Mueller Observatory features a 10 1/2-inch Warner and Swasey 155" focal length refractor built in 1899. The Observatory was constructed in 1960. The Cleveland Museum of Natural History also houses the Hanna Star Dome, a 4.9-meter copper dome capable of displaying 12 monthly skies. The dome was originally built in 1936 and illuminated by 3,000 lamps. Currently the dome is illuminated by one lamp, with stars displayed by the use of fiber optics. The Hanna Star Dome is a visitor controlled hands-on planetarium display.

The planetarium and observatory provide programming to public and private schools, elementary and secondary schools, local universities such as Case Western Reserve University, and the general public. School presentations are available throughout the

Photo: Cleveland Museum of Natural History

year and must be scheduled through the Education Division. Admission for school presentations varies depending on group size and presentation. The price per student ranges from $3.25 to $5.50, and nearly 17,000 students visit the planetarium each year.

Public programs at the Ralph Mueller Planetarium are presented on Saturdays and Sundays. Family programs are available at 1 p.m. (with an additional family program on Saturday at 11 a.m.) and all ages are welcome. Additional public programs are presented on Saturday and Sunday at 2, 3 and 4 p.m. These later afternoon programs are restricted to those of school age and above. Admission to the public programs is $1.50 above the normal Museum admission fee. All fees mentioned are subject to change. There are additional public programs during the months of June, July and August. These programs occur Monday through Friday at 2 p.m., with an additional family program offered on Wednesday at 11 a.m. Nearly 20,000 people attend public planetarium programming throughout the year. The Hanna Star Dome is available to museum visitors anytime the museum is open. The Ralph Mueller Observatory is open on Wednesday evenings at 8:30 p.m., September through May.

Museum of Scientific Discovery

Strawberry Square
3rd & Walnut Street
Mail:
P.O. Box 934
Harrisburg, PA 17108
Phone: (717) 233-7969
Fax: (717) 233-4888
E-mail: discover@msd.org
Internet: www.msd.org

The Skylights Astronomy Center at The Museum of Scientific Discovery projects the stars, moon and bright planets on the inside of an 18-foot diameter dome. A satellite dish on the museum's roof allows the center to bring live NASA images and other video programs to the Skylights Astronomy Center. Seasonal Sky Shows are available, as are special group programs and Topical Sky Shows. The Museum of Scientific Discovery also has a STARLAB portable planetarium, allowing them to bring the stars to virtually any location. Call 233-7969, extension 10, for details. The museum is closed Mondays and open until 5 p.m. Tuesday through Sunday (opening times vary).

NASA Headquarters

Public Services Division (POS)
300 E St. SW
Washington, DC 20546
NASA Headquarters: 202/358-0000
World Wide Web: www.nasa.gov

See separate entries for NASA facilities.

National Air and Space Museum

National Mall
7th and Independence Avenue, S.W.
Washington, D.C. 20560
World Wide Web: www.nasm.edu

> *This museum is a national treasure.*

Plan on spending the entire day at the National Air and Space Museum and you *might* have time to see everything you wanted. This museum is one of the finest in the nation and features an immense collection of space artifacts. For example, in Gallery 210 (Apollo to the Moon) you'll find the Apollo Lunar Roving Vehicle, Apollo Fuel Cell, Apollo Command Module Guidance Computer, Gemini 7 Spacecraft, Lunar Module Cockpit Mockup, Lunar samples, and much, much more. And this is only one of the many galleries at the National Air and Space Museum. You'll also want to check out Gallery 100 — *Milestones of Flight*, Gallery 110 — *Looking at Earth*, Gallery 207 — *Exploring the Planets*, Gallery 111 — *Stars*, Gallery 112 — *Lunar Exploration Vehicles*, Gallery 113 — *Rocketry and Space Flight*, and, of course, Gallery 114 — *Space Hall*. Guided tours are available.

The National Air and Space Museum also features the Albert Einstein Planetarium (see separate listing) and the Langley Theater, which projects IMAX films onto a screen seven stories wide and five stories high.

This museum is a national treasure. If you're visiting Washington D.C., don't miss it!

New York Hall of Science

47-01 111th Street
Flushing Meadows
Corona Park, NY 11368
Phone: (718) 669-0005
Fax: (718) 669-1341
Internet: www.nyhallsci.org

Photo: Ansell Horn

Queens Borough President Claire Shulman explores the planets on the new Window on the Universe exhibit, part of the new addition at the New York Hall of Science.

Ranked as one of the best science museums in the country, the New York Hall of Science is a leading innovator in exhibit technology and education programming. The *Window on the Universe* exhibit offers a vast menu of astronomical resources and viewing options, such as observing real-time images from on-line observatories around the world; operating the Hall of Science remote-control, on-line telescope (accessible on the World Wide Web); accessing the latest photos from the Hubble Space Telescope; or taking a simulated tour of the solar system using special computer programs.

Images of planets, stars, distant galaxies, and other astronomical subjects can be viewed on a video wall composed of 16 25-inch television monitors. Single images can occupy all 16 screens at once, or the screens can be divided into sections to view different images. Admission to the Hall of Science is $4.50 for adults, $3 for children (4-15) and senior citizens.

Noble Planetarium

Fort Worth Museum of Science and History
1501 Montgomery Street
Fort Worth, TX 76107-3079
Phone: (817) 732-1631
Fax: (817) 732-7635
Internet: 198.215.124.201/ fwmsh.html

Image: Noble Planetarium

The Fort Worth Museum of Science and History began as the Fort Worth Children's Museum in 1949 and featured a rough 18-foot dome with a Spitz Model A projector. The museum moved to its present location on Montgomery Street in 1954 and opened the Noble Planetarium the following year. The planetarium, named in honor of Miss Charlie Mary Noble, a local educator, features a Spitz A3P Prime projector, circular bench seating, and a 30-foot dome.

The Noble Planetarium includes a 360 degree projection gallery. Using special effects projectors, visitors are "transported" through space and time to explore the many aspects of the universe. More than 80,000 visitors enjoyed planetarium shows last year. Admission to the planetarium is $3 for all seats (in addition to the museum admission charge). Call for ticket information and show times.

The Museum also features an 80-foot domed OMNI Theater, eight Exhibit Galleries (including dinosaur exhibits), a variety of hands-on special exhibits, even a special place for visitors under the age of six. More than a million people visit the museum each year

Novins Planetarium

Director: Dr. R. Erik Zimmermann
Ocean County College
P.O. Box 2001
Toms River, NJ 08754-2001
Phone: (732) 255-0343
Show schedule: (732) 255-0342
Fax: (732) 255-0444

The Robert J. Novins Planetarium offers public programs on weekends (Friday through Sunday) throughout the year, and on weekdays during the summer. Their selection of Sky Shows include programs such as *Light Years from Andromeda, Tales of a Summer Night, Young Children's Show*, and *Concert Under the Stars*, among others. School programs are available on weekdays during the school year. The planetarium has a gift shop, an astronomy club (the Astronomical Society of the Toms River Area - ASTRA), and offers various astronomy courses.

Oakton High School Planetarium

Jack Steiffer, Planetarium Teacher
2900 Sutton Road
Oakton, VA 22180-2050
Phone: (703) 319-2735

The Oakton High School planetarium opened April of 1968 and features a Spitz A3P projection system, a 30-foot diameter dome, and seating for 84 (56 permanent and 28 folding chairs) in a chevron pattern.

Oklahoma Baptist University Planetarium

Kerry Magruder, Director
500 W. University, #61772
Shawnee, OK 74801
Phone: (405) 878-2090
E-mail: KVMagruder@aol.com
Internet: www.okbu.edu/ obu_nat_sci/planet/index.htm

Located in the W.P. Wood Science Building on the campus of Oklahoma Baptist University (OBU), the planetarium offers regular public showings Fridays at 7 and 8 p.m. while school is in session. Tickets for Friday night shows are $2 each. Group shows are available by appointment for a group charge of $25 (maximum seating capacity is 25).

Onizuka Space Center

Kona International Airport
Mail:
P.O. Box 833
Kailua-Kona, HI 96745
Phone: (808) 329-3441
Fax: (808) 326-9751

ASTRONAUT
ELLISON S. ONIZUKA
SPACE CENTER

The Ellison S. Onizuka Space Center is an educational facility dedicated to the memory of Hawaii's first astronaut. Colonel Onizuka perished on January 28, 1986 aboard the Space Shuttle Challenger, Mission 51-L.

The Onizuka Space Center presents a variety of exhibits, including *A Salute to the Challenger Astronauts*, *The Astronauts: Apollo to the Moon*, *The Mir Space Station*, and the *Space Shuttle* exhibit, plus a 45-seat space theater, a moon rock from the final moon landing (Apollo 17 in 1972), and more.

dedicated to the memory of Hawaii's first astronaut. Colonel Onizuka perished on January 28, 1986 aboard the Space Shuttle Challenger

Ott Planetarium

John Sohl, Director
2508 University Circle
Ogden, UT 84408-2508
Phone: (801) 626-6855
Internet: physics.weber.edu/planet/
ott.html

The Layton P. Ott Planetarium features a 30-foot hemispherical dome, a Spitz A4 star projector, and seating for 61. The planetarium is used for undergraduate astronomy classes, but also serves as a science education facility by providing programs to elementary students, secondary students, and the general public. For the general public, a program featuring some topic of current interest in astronomy is given every Wednesday night during the academic year. The Ott Planetarium also hosts The Ogden Astronomical Society, which supports the planetarium in many ways including the co-hosting of public star parties.

Owens Science Center

Russell Waugh, Planetarium Director
9601 Greenbelt Road
Lanham, MD 20706-3397
Phone: (301) 918-8750
Fax: (301) 918-8753
E-mail: rwaugh@umd5.umd.edu
Internet: www.gsfc.nasa.gov/
hbowens/hbowens_home.html

The Howard B. Owens Science Center is part of the Prince George's County School System and features a planetarium, the Challenger Learning Center, and a nature trail. The Challenger Learning Center offers a simulated space environment which includes a mission control center and a space station. Accommodating 16 to 32 students per "mission," the Challenger Learning Center gives students an opportunity to experience the technology and environment of space exploration. Two missions are flown each school day and opportunities are available for other paying groups to use the facility on Saturdays.

The Owens Science Center planetarium is open to the public Friday evenings at 7:30 p.m. when school is in session and hosts open house nights twice each school year. Call for show information.

The Challenger Learning Center offers a simulated space environment which includes a mission control center and a space station.

Palmer Observatory

**Santa Barbara Museum of
Natural History
2559 Puesta del Sol Road
Santa Barbara, CA 93105-2936
Phone: (805) 682-4711
Fax: (805) 569-3170**

The Palmer Observatory is located on the grounds of the Santa Barbara Museum of Natural History (see Gladwin Planetarium). The observatory is used for public education both by the Santa Barbara City College, which uses the observatory for its lab classes, and by the Astronomical Units, the museum's astronomy club, which conducts public observing sessions at the observatory about once a month.

Paulucci Space Theater

**Hibbing Community College
1502 E. 23rd Street
Hibbing, MN 55746
Phone: (218) 262-6720
Fax: (218) 262-6719
E-mail: P.Davis@HI.CC.MN.US**

Paulucci Space Theater at Hibbing Community College produces and presents programs for both school groups and the general public. The facility features a 60-foot dome, Spitz star system, 870 projector for films, 25 different slide projectors, and numerous special effects for multi-media presentations. This results in a spectacular presentation. While the majority of shows are astronomy related, some are for purely entertainment purposes.

Peninsula Planetarium and Abbitt Observatory

**Virginia Living Museum
David Maness, Director of Astronomy
524 J. Clyde Morris Blvd.
Newport News, VA 23601
Phone: (757) 595-1900
Fax: (757) 599-4897
E-mail: pegasus321@AOL.com
Internet: users.visi.net/~stargazr**

The Peninsula Planetarium features a 9.1-meter dome, a Spitz Model A3P projector, and seating for 70 in a unidirectional orientation. The planetarium is part of the Virginia Living Museum which features live native animals in natural settings. School programs are performed Monday through Friday by reservation. Public programs are offered at 3:30 p.m. daily and at 11 a.m., 1:30, 2:30, and 3:30 p.m. Saturdays, and 1:30, 2:30, and 3:30 p.m. on Sundays. Admission to public programs is $2.50 (or $1.00 extra when combined with Museum admission.) The planetarium office is open from 9 a.m. to 5 p.m.

The Museum's Abbitt Observatory features a Celestron 14-inch telescope with a hydrogen alpha rejection filter. It is open daily for solar observations, and Thursday evenings from 7 p.m. until 9 p.m., weather permitting.

Perkins Observatory

P.O. Box 449
Delaware, OH 43015
Phone: (614) 363-1257

Shortly before his death in 1924, Professor Hiram Mills Perkins donated $200,000 to Ohio Wesleyan University to build an observatory with a telescope larger than all others in the country at that time, with the exception of the 100-inch Mt. Wilson telescope. On May 23, 1923, Professor Perkins, who was 90 years old, turned the first sod at the observatory's ground-breaking ceremony. Professor Perkins' vision went beyond the observatory itself, and included an opulent entrance, star-lined foyer ceiling, a 100-seat lecture hall, a well-stocked library, a kitchen, a darkroom, and thousands of square feet of office space (now being converted for museum displays and a bedroom for weary astronomers).

In March 1931, Perkins Observatory began publishing a quarterly review called *The Telescope*, which they continued publishing until it became part of the prominent *Sky & Telescope* magazine.

The observatory hosts stargazing nights on most Friday and Saturday evenings (except holiday weekends). The programs include planetarium shows, tours of the observatory, a slide show on beginning astronomy, and observing with the 32-inch telescope, weather permitting. Programs begin at 8 p.m. (9 p.m. May - September). Tickets can be ordered in advance, or purchased at the door (though most programs are full and at-the-door tickets are not available).

Daytime programs for groups can be reserved for a nonrefundable fee of $100 per presentation, plus a $1 fee for each person who attends. Night programs can also be reserved for groups for a nonrefundable free of $150, plus a $1 fee for each person attending.

The Columbus Astronomical Society meets at the observatory for its regular club meetings on the second Saturday of each month at 8 p.m. The club sponsors some of the programs at Perkins.

Photo: Ohio Wesleyan University

This historic picture of the interior of Perkins Observatory appeared in Le Bijou *in 1927*

The observatory hosts stargazing nights on most Friday and Saturday evenings

Pine Mountain Observatory

P.O. Box 5795
Eugene, OR 97405
Phone: (541) 382-8331
Internet: pmo-sun.nero.net

Located 30 miles southeast of Bend, Oregon, Pine Mountain Observatory is one of the few observatories to offer access to the public for nighttime observing on a regular basis, and has some of the largest telescopes in the Pacific Northwest. Visitors get to experience a "working observatory" rather than a tourist facility or display museum. The skies are usually very clear and dark and the Milky Way is so bright that it is often mistaken for a cloud.

The facility, owned by the University of Oregon, features 24-inch and 15-inch Cassegrain telescopes, "big eyes" binoculars, and is open for public drop-in visits on Friday and Saturday evenings, mid-May through late September. Tours commence at dusk (usually 9 p.m. in mid-summer) and feature a lecture about general astronomy and modern observing technologies, followed by observation of the Moon, Planets, and Deep Sky objects (weather permitting). Wide-field and/or piggybacked CCD cameras may be available, allowing visitors to create digitized images. Tours are conducted by experienced volunteer amateur astronomers who are members of the citizens' support group, Friends of Pine Mountain Observatory.

Visitors should come prepared for below-freezing weather (elevation 6400 feet), mountain/forest terrain underfoot, and bring small flashlights shielded with red or brown paper. $2.00 suggested donation per adult. Children are welcome, but infants probably will not be comfortable. Access may be restricted if research is going on, and nights near Full Moon phase will be too bright to see "deep sky" objects. For a virtual tour, visit the Pine Mountain Observatory web site.

Directions: From Bend, take State Highway 20 east, 26 miles, to Millican. Just past the Millican store, look for a turnoff to the right, south, through a cattle guard, onto a dirt road. Take this road 8 miles to the top of Pine Mountain. Stay on the main dirt road at all junctions. Road is "washboardy" and dusty in the summer, has hairpin turns, and is steep in places, regular vehicles can make the trip OK, but trailers and busses are NOT recommended. Allow one hour travel time from Bend. The closest motels/restaurants are in Bend, and fuel is only available at Millican during daytime hours. Although there are "porta-potties" at the Observatory, no water is available.

> *The facility, owned by the University of Oregon, features 24-inch and 15-inch Cassegrain telescopes, "big eyes" binoculars, and is open for public drop-in visits on Friday and Saturday evenings, mid-May through late September.*

The Planetarium at Santa Fe Community College

6401 Richards Ave.
Santa Fe, NM 87505
Phone: (505) 438-1777
Fax: (505) 438-1302
E-mail: SCHIPPI@santa-fe.cc.nm.us

Located in the upper level of the college's west wing, the Planetarium offers regular shows the first Thursday of the month at 7 p.m., Fridays at 6:30 and 8 p.m., and family programs on Saturdays at 10:30 a.m. Tickets are $3.50 for adults, $2 for children 12 and under. Tickets go on sale 30 minutes before show time and seating is limited. Call for show information.

Pomeroy Planetarium

Joe Guenter, Director
University of Arkansas - Monticello
Division of Math and Sciences
Monticello, AR 71656
Phone: (501) 460-1016
Internet: cotton.uamont.edu/
~guenterj/planet.html

Established in 1975 with a matching grant from the National Science Foundation, the Pomeroy Planetarium is located in the Turner Neal Museum of Natural History on the University of Arkansas-Monticello campus. In addition to the numerous programs presented , the planetarium offers Public Night Programs every month during the academic year. When weather permits, these programs are followed by telescope observing sessions. All events are open to the public and free of charge.

Raritan Valley Community College Planetarium

P.O. Box 3300
Somerville, NJ 08876-1265
Phone: (908) 231-8805
Fax: (908) 526-7938
Internet: www.raritanval.edu/
planetarium/index.shtml

The Raritan Valley Community College (RVCC) Planetarium presents shows (including laser shows) for the general public and for groups by appointment. A Starlab Workshop is offered during the summer months, allowing participants to enhance their knowledge through discussions and hands-on projects.

Rauch Memorial Planetarium

J. Scott Miller, Program Coordinator
University of Louisville
Louisville, KY 40292
Phone: (502)852-6664
FAX: (502)852-0831
E-mail: jsmill0l@homer.louisville.edu
Internet: www.louisville.edu/
planetarium

The Rauch Memorial Planetarium opened in 1962 and was Kentucky's first planetarium. The planetarium features a Spitz 512 projector, a 30-foot dome, and concentric, continuous seating for approximately 80 adults or 100 children. Multi-media programs, which use a host of auxiliary projectors, slide projectors, video, and computer imagery in support of the Spitz 512, are presented on a variety of topics in astronomy. Programs are presented for groups as well as the general public. School/group performances are offered during the week by reservation, while programs for the general public are on Saturday afternoons. The planetarium normally presents an evening program on the first Saturday of each month, except in May. Admission is $3.50 for adults (13 years and up) and $2.50 for children (4 - 12 years), senior citizens (60+), or University of Louisville/Metroversity college stu-

Photo: J. Scott Miller

dents with their identification. Group rates are $2.00 per person when a reservation is made. The planetarium also features a gift shop/bookstore located within the building.

Reuben H. Fleet Space Theater & Science Center

Dennis Mammana, Resident Astronomer
P.O. Box 33303
San Diego, CA 92163-3303
Phone: (619) 238-1233
Advance Ticket Office: (619) 232-6866
TDD Line: (619) 238-2480
Fax: (619) 234-4627
E-mail: mammana@rhfleet.org
Internet: www.rhfleet.org

The Fleet Space Theater features a 76-foot dome tilted down at a 25-degree angle in front of the spectators (who are all seated in tiered rows facing forward), giving the illusion of being suspended in space. Using the world's largest motion picture format, OMNIMAX® films are projected onto a dome, creating a super-realistic cinematic experience. As an added feature, the 3:00 p.m. and 4:00 p.m. shows daily are open captioned with subtitles for the hearing-impaired.

Robinson Observatory

Dr. Nadine G. Barlow, Director
Department of Physics
University of Central Florida
Orlando, FL 32816
Phone: (407) 823-2805
Fax: (407) 823-5112
E-mail: ngb@physics.ucf.edu

The University of Central Florida's (UCF) Robinson Observatory is located on the UCF campus, on the northeastern side of the city of Orlando. The observatory was made possible by a generous gift by Herbert and Susan Robinson of Orlando. The heart of the Robinson Observatory is its 0.65-m (26") Schmidt-Cassegrain reflecting telescope. The 4-ton telescope was built by Tinsley Laboratories in 1967-68 for the astronomy department at the University of Southern Florida (USF) in Tampa. After the USF astronomy program merged with that of the University of Florida (UF) in Gainesville, the telescope was put into storage until 1992. At that time, the Central Florida

Astronomical Society (CFAS) began working with UF and UCF to bring the telescope to Orlando and restore it to its former glory. The Robinson Observatory with its newly refurbished 0.65-m telescope was dedicated on April 25, 1996. The telescope is used by UCF faculty and students, UF astronomers, and members of the CFAS for astronomical research, public viewing opportunities, and educational outreach.

Tours of the observatory by pre-college school groups, members of interested organizations, and other special groups can be arranged by contacting the Observatory Director. The Robinson Observatory is open for public observing sessions twice each month and for special events. Call for dates, times, and directions.

Rock Creek Park Planetarium

Superintendent, Rock Creek Park
3545 Williamsburg Lane, NW
Washington, DC 20008-1207
Headquarters: (202) 282-1063
Planetarium: (202) 426-6829
Nature Center: (202) 426-6829
Internet: www.nps.gov/rocr/
planetarium/

Located at the Rock Creek Nature Center, the Rock Creek Park Planetarium serves as an astronomy laboratory, allowing visitors to study the sky under ideal conditions, and is the only planetarium in the national park system. The majority of Rock Creek's planetarium programs show the night sky as it appears in the Washington, D.C. area for the specific date and time of the program. The projector can also be accelerated to allow visitors to witness phenomena which take months, years, or even centuries to occur. Programs are usually 45-60 minutes in length.

The planetarium holds regular shows on Saturdays and Sundays at 1 p.m. for children four and older (4-7 year olds must be accompanied by an adult) and at 4 p.m. for children seven and older. There is also an after-school show at 3:45 p.m. every Wednesday, year round.

Rollins Planetarium

Young Harris College
1 College Road
Young Harris, GA 30582
Phone: (706) 379-4312
Fax: (706) 379-4306
E-mail: kentm@yhc.edu

Rollins Planetarium is located in the picturesque Appalachians of North Georgia. Wayne and Grace Rollins provided a grant to build the planetarium in 1978, and their generous philanthropy in 1995 provided many needed renovations. The planetarium contains a 40-foot dome, a Spitz System 512 projector, and will seat 111 people. Because of the Rollins' generous donations, there is never any charge for public shows or cosmic concerts. The Rollins' Planetarium provides the community with new shows every other month, and also offers educational shows to elementary and secondary schools in the area. Special shows during the year include cosmic concerts in October and May, and a Christmas show during November and December. Cosmic concerts are presented by the student organization, Sky Club, and are a compilation of music, spinning stars, and state of the art visual effects. Shows start promptly at 8 p.m. and, weather permitting, telescopes will be available for stargazing following the show.

Rosicrucian Egyptian Museum and Planetarium

Julie Scott, Director
Rosicrucian Park
1342 Naglee Avenue
San Jose, CA 95191
Phone: (408) 947-3635
Fax: (408) 947-3638
Internet: www.rosicrucian.org

Photo: Rosicrucian Egyptian Planetarium

The Rosicrucian Planetarium offers a program to introduce visitors to the wonders of the universe. The theater seats 100 people and the show lasts approximately 45 minutes. Take a trip back 5,000 years to the age when the faint star, Thuban, was the north star instead of our current Polaris. This program details astronomy's influence on Egyptian religion and myths—and shows how the ancient Egyptians used astronomy to build the Great Pyramid. School programs are presented Monday through Friday (except Tuesdays) by reservation only. Admission is $2.00 per student. Programs open to the general public are presented Monday through Friday (except Tuesdays) if the school groups don't take all of the available space. On Saturdays and Sundays, the Planetarium show begins at 2 p.m. Admission to public programs is $4.00 for adults, $3.50 for senior citizens and students and $3.00 for children under 15. The planetarium show is not suitable for children under 6.

> *Take a trip back 5,000 years to the age when the faint star, Thuban, was the north star instead of our current Polaris*

Saint Cloud State University Observatory and Planetarium

720 4th Avenue S., MS 324
St. Cloud, MN 56301-4498
Phone: (320) 255-2011
Fax: (320) 255-4262
E-mail: mnook@stcloudstate.edu
Internet: condor.stcloud.msus.edu/
~mnook/Planetarium.html

Located in the Math Science Center at St. Cloud State University (SCSU), the planetarium opened in 1973. There are now 80+ shows presented to over 4,000 people each year. Planetarium shows incorporate multimedia and group participation into an exciting learning experience. Young people are encouraged to view science as a worthwhile endeavor and a significant part of our culture. Call for show times.

San Diego Aerospace Museum

2001 Pan American Plaza
Balboa Park
San Diego, CA 92101
Phone: (619) 234-8291
Internet: www.aerospacemuseum.org

Photo: San Diego Aerospace Museum

Over 65 U.S. and foreign aircraft and spacecraft are on display at the San Diego Aerospace Museum, including a Mercury Spacecraft, Gemini Spacecraft, and Apollo Spacecraft. Displays allow you to explore aviation history from the Wright Flyer to the Space Shuttle. Open daily, except Thanksgiving and Christmas. Active duty military are admitted free.

Santa Rosa Junior College Planetarium

Ron Oriti, Planetarium Director
1501 Mendocino Avenue
Santa Rosa, CA 95401
Phone: (707) 527-4371
Fax: (707) 527-4870

a variety of programs designed for K-12 students and the general public

Located in room 2001 of Lark Hall, the Santa Rosa Junior College (SRJC) Planetarium features a 40-foot dome, seating for 100, and a Spitz-Goto model SG12 prime sky projector. The planetarium serves as an instructional facility for students of the college, but also has a variety of programs designed for K-12 students and the general public. Up to 12 school shows are presented each week during the school

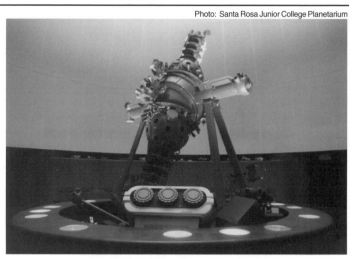

Photo: Santa Rosa Junior College Planetarium

Spitz-Goto model SG12 prime sky projector

year (call for reservations). Admission is $1 per person. Six public shows are presented each week during the regular school year at $4 for general admission, $2 for students and seniors.

Sargent Space Theater

Science Center of Iowa
David Elliot, Director
Kristian Anderson, Operations
Steve Cooper, Production & Technical
Harry Aller, Laserist
4500 Grand Avenue
Des Moines, IA 50312-2499
Phone: (515) 274-6868
E-mail: schedule@sciowa.org
Internet: www.sciowa.org

The Sargent Space Theater features a 40-foot dome and unidirectional seating for 125. Equipment includes a Digistar I, Barco 1109 video projector, and a Laser Fantasy E3500 laser projection system. The Sargent Space Theater provides astronomy education and entertainment for public, civic and school groups. Group reservation rates are available. For the general public, planetarium admission is free with Science Center admission, which is currently $5 for adults and $3.50 for ages 3-12 and senior citizens.

Laser Rock concerts are offered on Friday and Saturday nights and cost $5.75, while Laser Matinees are offered on Saturdays and Sundays for $1.50. Call for current show times. The Science Center of Iowa also houses a number of exhibit galleries, a Challenger Learning Center, and a gift shop. The planetarium office is open Monday through Friday from 9 a.m. to 5 p.m. for your questions and reservations.

> *The Science Center of Iowa also houses a number of exhibit galleries, a Challenger Learning Center, and a gift shop*

Schenectady Museum and Planetarium

Urban Cultural Park Visitor Center
Nott Terrace Heights
Schenectady, NY 12308
Phone: (518) 382-7890

The Schenectady Museum and Planetarium (SM&P) is a private, not-for-profit organization whose goal is science education. Annual attendance exceeds 50,000 with over 60% of that figure attending its planetarium. The SM&P is the only member of the Association of Science and Technology Centers (ASTC) in its region and it houses the Schenectday Urban Cultural Park, part of the New York State (NYS) cultural park system.

The SM&P is occupied by the only professionally operated public planetarium (SMP) in the northeastern quadrant of NYS and is dedicated to teaching astronomy and bringing astronomy information to the public. Programming consists of 12 grade-level school shows plus rotating public programs. Planetarium hardware includes a Spitz A3PR star projector and console, 15 Kodiak Ektagraphic projectors and 12 special effect projects.

Seven Hills Observatory

Mark Urwiller, Director
4711 Heather Lane
Kearney, NE 68847
Phone: (308) 234-6536
E-mail:
murwille@genie.esu10.k12.ne.us
Internet: genie.esu10.k12.ne.us/
~murwille/7hills.htm

The Seven Hills Observatory was built in the Summer of 1996 in the Seven Hills Subdivision, approx. 4 miles Northwest of downtown Kearney Nebraska. The Observatory is 20' x 20' with a roll off roof and was constructed almost entirely of low thermal mass materials. Call or write for details on public viewings, star parties, and other events.

Sherzer Observatory

Norbert Vance, Director
Eastern Michigan University
Ypsilanti, MI 48197
Phone: (313) 487-4146
Fax: (313) 487-0989
E-mail: phy_vance@online.emich.edu
Internet: www.emich.edu/public/
art_sci/phy_ast/sherzer.htm

Located on the top floor of historic Sherzer Hall, Sherzer Observatory was entirely reconstructed after a devastating fire in 1989. The updated facility includes a 10-inch apochromatic refractor telescope with ancillary equipment, numerous portable telescopes, observing deck, laboratory classroom with planetarium, instructional computer network with Internet access, museum room, and darkroom. The observatory supports department introductory and observational astronomy courses and serves as a base for the Eastern Michigan University (EMU) Astronomy Club (see separate listing under *Organizations*). The EMU Astronomy Club sponsors open houses and observing sessions throughout the academic year (reduced summer schedule). Sessions are free and open to students and the general public on clear Monday evenings, 7:30 p.m. to 10 p.m., and Thursday evenings, 8:30 p.m. to 10 p.m. School and activity groups can arrange for tours by appointment.

Silverman Planetarium

Milton J. Rubenstein Museum of
Science & Technology
500 S. Franklin Street
Syracuse, NY 13202
Phone: (315) 425-9068
Fax: (315) 425-9072

Photo: Rubenstein Museum

The Rubenstein Museum of Science & Technology (MOST) offers three levels of science in the form of hundreds of exhibits, the Bristol Omnitheater, and, of course, the Dr. & Mrs. Herbert J. Silverman Planetarium. Three shows are offered daily under the planetarium's 24-foot domed star theater, and though the regular planetarium shows are not recommended for children under 4 years of age, a special program called "A Zoo in the Sky" is presented for these younger visitors at 10:30 a.m. on weekends and special holidays. Regular planetarium shows are at 11:30 a.m., 2 p.m., and 4 p.m. Planetarium tickets are $.50, in addition to regular admission. Members enter free.

Smith Planetarium

Jim Greenhouse, Planetarium Curator
The Museum of Arts and Sciences
4182 Forsyth Road
Macon, GA 31210-4806
Phone: (912) 477-3232
Fax: (912) 477-3251

The Mark Smith Planetarium features a 40-foot dome, seating for 118, and a Minolta MS-10 star projector that displays about 4,500 stars. The MS-10 is supported by three video projectors, several special effects projectors, and about 30 computer controlled slide projectors arrayed to project on various parts of the dome. Public planetarium programs are presented at 4 p.m. daily with additional shows on weekends and holidays.

The Museum of Arts and Sciences also has an observatory with two 17.5-inch Dobsonians, two equatorially mounted Schmidt-Cassegrains on permanent pillars (one 14-inch and one 10-inch), and two 8-inch portable Schmidt-Cassegrains. The observatory is open to the public every Friday night when the sky is clear. Special programs are traditionally offered to celebrate Astronomy Day, the Perseid Meteor Shower, and other astronomical phenomena. The planetarium and observatory are included with admission to the museum.

Smith Planetarium
Pacific Science Center
Seattle, WA
Phone: (206) 443-2001
Office: (206) 443-3648
Astronomy Hotline: (206) 443-2920
Internet: www.pacsci.org

Located within Pacific Science Center, the Willard W. Smith Planetarium seats approximately 40 people and is small enough to encourage interaction between visitors and the Planetarium Demonstrator. All demonstrations are live, and visitors are welcome to ask questions throughout the show.

Planetarium demonstrations are free and open to the public every day. School groups and other private groups can also reserve the Planetarium for a customized show. Planetarium visitors must be four years old or older. There is no late seating.

Pacific Science Center is located in Seattle, Washington, just steps away from the Space Needle. Look for the five 85-foot white arches to guide your way.

Sommers-Bausch Observatory
University of Colorado
Boulder, CO 80309
Phone: (303) 492-6732

In 1946 the Bausch & Lomb Company gave the University of Colorado a 10.5-inch refractor, formerly the property of Carl L. Bausch. The telescope was designed by George Saegmuller, a B&L designer, in 1912. Housing for the telescope became possible when, in 1949, the University of Colorado received a bequest of $49,054 from the estate of Mrs. Mayme Sommers in memory of her husband Elmer E. Sommers. These funds were to be used to construct an observatory building. The new Sommers-Bausch Observatory was dedicated on August 27, 1953, during the 89th meeting of the American Astronomical Society.

In 1971 the University received a $164,000 grant from the Science Development Program of the National Science Foundation for the purchase of a 24-inch Cassegrain-coude reflector. The telescope was installed in 1973 in the original SBO dome. A 4,000 square foot addition was built onto the observatory in 1981, providing four darkrooms (the original darkroom being converted into a reading room), a laboratory, several offices, a workshop, and a large open observing deck. In 1980 the National Science Foundation made an additional grant of $135,000 for the purchase of an 18-inch Cassegrain reflector.

Fiske Planetarium is also located on the University campus (see separate listing).

Southworth Planetarium

University of Southern Maine
96 Falmouth Street
Mail:
P.O. Box 9300
Portland, ME 04104-9300
Phone: (207) 780-4249
Fax: (207) 780-4051
E-mail:
deines@portland.caps.maine.edu
Internet: www.usm.maine.edu/
~planet/

The Southworth Planetarium seats 63 and is open to the public for Friday and Saturday evening shows and Saturday matinees. From October through April, they offer Sunday matinees. Group shows can be arranged by appointment. A variety of sky shows and holiday special shows are available. Music laser light shows are also available. Call for details and show times.

Space Center Houston

1601 NASA Road 1
Houston, TX 77058
Phone: (281) 244-2100
Fax: (281) 283-7724

Space Center Houston is the official visitors center for NASA's Johnson Space Center. This world-class facility is not federally funded, but is owned and operated year-round by the Manned Space Flight Education Foundation, Inc., a not-for-profit corporation formed to educate and inspire the public about the space program.

Attractions at Space Center Houston are self-guided so you can spend as much time, or as little time, in each area as desired. The Space Center Theater has a 5-story movie screen where you can watch IMAX films that will take you on a journey to outer space.

Other attractions allow you to learn what it would be like living and working in space (the 20-minute *Living in Space* presentation); test your piloting skills by landing an 80-ton orbiter via simulation or retrieving a satellite with the Manned Maneuvering Unit. Try on a real space helmet; and experience the excitement of space in the hands-on Kids Space Place.

A NASA Tram Tour is available which will take you behind the scenes at Johnson Space Center. You'll have the opportunity to see Mission Control and the astronaut training facilities, including the Space Environment Simulation Laboratory and the Mock-Up & Integration Laboratory. Tours vary so check with a crew member about the areas you will visit.

A changing variety of special exhibits and events are part of the Space Center Houston experience. A mix of space themes are represented, from stellar art exhibits to cosmic photographs of space.

Membership in Space Center Houston is available and allows you to support the facility and enjoy its benefits throughout the year. Members are entitled to unlimited visits during regular operating hours, this includes IMAX films and special exhibits. Plus members receive a discount at the SpaceTrader Gift Shop and Zero-G Diner. Members pay no parking fee and receive a bimonthly newsletter.

SpaceQuest® Planetarium

The Children's Museum of Indianapolis
3000 North Meridian Street
Indianapolis, IN 46208-4716
Mail:
P.O. Box 3000
Indianapolis, IN 46206-3000
Phone: (317) 924-5431
Fax: (317) 921-4019
E-mail: tcmi@childrensmuseum.org
Internet: www.childrensmuseum.org

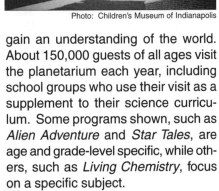

Photo: Children's Museum of Indianapolis

The SpaceQuest Planetarium is one of the main attractions at The Children's Museum of Indianapolis. Dedicated to enriching the lives of children and adults alike, the museum offers educational and entertaining experiences that challenge visitors to ask questions, use their imagination and gain an understanding of the world. About 150,000 guests of all ages visit the planetarium each year, including school groups who use their visit as a supplement to their science curriculum. Some programs shown, such as *Alien Adventure* and *Star Tales*, are age and grade-level specific, while others, such as *Living Chemistry*, focus on a specific subject.

The planetarium offers a unique *Laser Fantasy* show. This program presents laser and Digistar effects set to music. For those wanting to learn more about the night sky the planetarium offers *Starwatch*, a live, interactive presentation highlighting the night sky and other current events in astronomy every first Thursday of the month.

Photo: Children's Museum of Indianapolis

Staerkel Planetarium

Parkland College
2400 West Bradley Avenue
Champaign, IL 61821-1899
Phone: (217) 351-2568
Show Information: (217) 351-2446
Fax: (217) 351-2581
E-mail: dleake@parkland.cc.il.us
Internet: www.parkland.cc.il.us/
coned/pla/staerkel.html

The William M. Staerkel Planetarium at Parkland College is the second largest planetarium in Illinois. The planetarium opened in the fall of 1987 and boasts a Carl Zeiss M1015 Star Projector, the first of its kind in the world, which provides 7,600 stars, 5 planets, the sun and moon, plus numerous nebulae, star clusters, and galaxies. These objects are projected onto a 50-foot aluminum dome suspended from the ceiling. Beneath the dome are 144 seats. In addition to the star projector, the planetarium offers a full range of video projection, 35mm film projection, nearly 60 slide projectors under computer control, a host of special visual effects, and a 3,000 watt sound system. The planetarium lobby displays "Cosmic Blink," a mural by Billy Morrow Jackson done especially for the planetarium.

The Staerkel Planetarium is open throughout the week for schools and civic groups. Nearly three dozen programs are available for teachers from preschool to college (a teacher's brochure may be obtained by calling the planetarium office). Public shows are offered almost every Friday and Saturday evening. These shows include live tours of the current night sky, 35mm films, automated features, and rock-n-roll light shows.

The planetarium is also active in community outreach programs, including a presentation that features a simulated space shuttle space suit. There is a membership organization called the "Friends of the Staerkel Planetarium" where participants are entitled to free shows and a quarterly newsletter.

Stanback Planetarium

South Carolina State University
300 College Ave.
Orangeburg, SC 29117
Mail:
P.O. Box 7636
Orangeburg, SC 29117
Phone: (803) 536-8711
E-mail: starman@scsu.edu
Internet: www.draco.scsu.edu/

Public Shows are given on the second Sunday of each month at 3 p.m. and 4 p.m. Stanback Planetarium offers school shows Mondays through Thursdays at 9 a.m., 10 a.m. and 11 a.m. Call (803) 536-7174 for reservations.

The NASA Regional Educator Resource Center is also located at Stanback Planetarium and can be reached at (803) 536-8709.

Starwalk Planetarium

Henry Bouchelle, Director
Colonial School District
Chase Avenue
New Castle, DE 19720
Phone: (302) 429-4013
Fax: (302) 429-4005
E-mail: hbouchel@udel.edu

The Colonial School District in New Castle, Delaware boasts the largest planetarium in the state, with seating for 60. Its Spitz A4 projector creates a realistic starfield of 1,854 stars, nebulae, and galaxies on the 30-foot dome.

Planetarium lessons are a regular part of the district's science program at the primary, intermediate, and middle school levels. Each year the planetarium hosts approximately 14,000 students from every school in the Colonial School District.

Starwalk Planetarium is home to the earth/space science program called Project Starwalk. This program was conceived, developed, nationally validated, and originally disseminated at the planetarium. Project Starwalk has been adopted and further disseminated by school districts across the nation and active development of the program continues at the planetarium.

In addition to serving the school district, Starwalk Planetarium is a useful community resource, hosting numerous meetings, teacher workshops, and civil groups, including the Delaware Astronomical Society. The planetarium's director, Mr. Henry Bouchelle, has an impressive list of qualifications, including his service as assistant director of Project SPICA and as lead instructor for the American Astronomical Society's AASTRA programs. In 1985 he was selected as a NASA Teacher in Space finalist.

Stennis Space Center

Public Affairs Office
Stennis Space Center, MS 39529
Phone: (601) 688-3341
Visitor Center: (601) 688-2370
or 1-800-237-1821
TRC: (601) 688-3338
E-mail: pao@ssc.nasa.gov
Internet: www.ssc.nasa.gov

NASA's John C. Stennis Space Center (SSC) in South Mississippi has become one of the Mississippi Gulf Coast's leading tourist attractions. More than 100,000 people from all over the world visit the space center each year, representing all 50 states and 60 foreign countries.

Stennis' Visitors Center offers exciting and informative programs for visitors of all ages. Guided tours, films, demonstrations and an extensive collection of exhibits and displays highlight the visit to SSC.

The guided tour takes visitors through the Space Shuttle Main Engine test complex. This is where the shuttle's powerful main engines are proven flight worthy before being placed on the orbiter. Guests have the opportunity to view an engine test if one occurs during Visitors Center hours of operation. The tour also takes visitors past Stennis' hyacinth pond, a natural water treatment facility, and by the 7 1/2-mile, man-made canal.

Outdoor exhibits at the Visitors Center include: a Space Shuttle external fuel tank and solid rocket booster; a scale model of a Saturn V rocket, which took America's astronauts to the moon; a 1/25-scale model of the Space Shuttle; and a full-scale Nomad buoy from the National Data Buoy Center, which is used to collect data in ocean areas that experience hostile weather environments. Also on display is an F-1 engine that powered the first stage of the Saturn rocket and a 68-foot Jupiter-C rocket, sister ship to the vehicle that carried America's first satellite into orbit.

In the Visitors Center auditorium, guests enjoy a wide selection of movies related to space flight, living and working in space, and future space projects. The indoor display area contains a moon rock estimated to be over three billion years old, which was collected by the Apollo 15 mission crew. An exhibit featuring astronauts from Mississippi and Louisiana and a mural from the Paris Air Show can also be seen. Space Believe is a new interactive, hands-on science learning center where children can use games and projects to accomplish tasks just like astronauts perform in space. Visitors may also surf the Internet to find out more information on a wide variety of subjects ranging from space ex-ploration to materials available at Stennis for teachers.

Stennis Space Center's Visitors Center is open seven days a week, except Easter, Thanksgiving and Christmas. Hours are 9 a.m. to 4 p.m. Monday through Saturday and noon to 4 p.m. on Sunday. There is no admission charge. Tours depart on a regular basis each day, and cameras are allowed throughout the Visitors Center and on the guided tour. Special presentations for groups can be arranged by contacting the Visitors Center for reservations.

The SSC Teacher Resource Center (TRC), located in the Visitors Center, has a large collection of materials for educators, including hundreds of videotapes, computer software, printed materials and lesson plans, which reflect the most recent scientific discoveries about space, oceans and the Earth. Subjects range from science and geography to meteorology, astronomy, social studies and environmental science. The Teacher Resource Center has served more than 50,000 educators primarily from Mississippi and Louisiana since opening in 1985.

Stennis Space Center is located on Mississippi Highway 607 with easy access from Interstates 10 and 59. Signs are located along the roadways with directions to the space center and the Visitors Center.

Space Believe is a new interactive, hands-on science learning center where children can use games and projects to accomplish tasks just like astronauts perform in space

Stockton College Observatory

Hal Taylor, NAMS
Stockton College of New Jersey
Pomona, NJ 08240-0195
Phone: (609) 652-4471
Fax: (609) 748-5515
E-mail: hal@astro.stockton.edu
Radio: K2PT
Monitor 21.430 MHZ

The Stockton College Observatory is a 16-foot ash dome housing a 14-inch Celestron and other small instruments. It is located on the campus of the Richard Stockton College of New Jersey, about 12 miles northwest of Atlantic City. Public viewing sessions are held on clear Friday nights from September through April, from 8-10 p.m. Sessions held between May and August are 9-11 p.m. on the first and third Fridays (when clear).

Strickler Planetarium

Olivet Nazarene University
P.O. Box 592
Kankakee, IL 60901-0592
Phone: (815) 939-5308
Fax: (815) 939-5071
E-mail: bschroed@olivet.edu
Internet: www.olivet.edu/
departments/planet

Brock Schroeder, Director of Strickler Planetarium, invites you to explore the universe and learn about the stars, comets, and the planets. Special shows are available during the holiday seasons. Call for monthly public show times. The observatory is also open for tours on a limited basis. Call for details.

Planetarium Admission: $1

Stull Observatory

Dave Toot, Director
Alfred University
Alfred, NY 14802
Phone: (607) 871-2208
E-mail: ftoot@bigvax.alfred.edu
Internet: www.alfred.edu

With an impressive collection of five major telescopes, the Stull Observatory serves first and foremost as a resource for basic science education of undergraduate students of all majors and backgrounds. It is a crucial component of the Astronomy Minor program (part of the Physical Sciences Division at Alfred University). The observatory also serves the community at large through public open houses and via teachers workshops and school group tours. Telescopes include a 9-inch Henry Fitz refractor, a 16-inch Cassegrain reflector, and 14, 20 and 32-inch Newtonian reflectors. The 32-inch telescope is computer controlled.

Photo: Stull Observatory

An overhead view of Stull Observatory

Sudekum Planetarium

Cumberland Science Museum
800 Fort Negley Blvd.
Nashville, TN 37203
Phone: (615) 401-5077
Information: (615) 862-5160
Astroline: (615) 401-5092
Reservations: (615) 862-5177
Fax: (615) 401-5086
E-mail: mccallk@ten-
nash.ten.k12.tn.us

The Sudekum Planetarium is typically considered to be a mid-size planetarium, housing a Spitz 512 star projector under a 12.2 meter/40-foot dome. While there is a lot of auxiliary equipment to help make shows more dramatic or add a little flash, it all comes together because of the dynamic education staff. The most popular programs feature the stars and planets that can be seen in the current night sky or a trip to explore the planets in our solar system. Other topics range from the birth of stars to the end of the universe. The Planetarium also has a STARLAB portable planetarium which travels throughout Middle Tennessee on an all-star tour.

In addition to regularly scheduled programs, the Planetarium works with the Barnard-Seyfert Astronomical Society and other local organizations on projects such as star parties and astronomy day. Free public star parties are held throughout the year at several locations around the area.

The staff of the Sudekum Planetarium has been producing complete original programs for almost ten years. Once a program premieres in Nashville, the show can be duplicated and is made available for purchase by other planetariums across the country and around the world. By the spring of 1997, almost three hundred shows had been sold in eight years.

Sudekum-produced stellar experiences include *The Planet Patrol: Solar System Stake Out*, *Our Place In Space*, *Just Imagine*, and others. Many of these productions have won awards. *The Light-Hearted Astronomer* received second place in the 1990 Eugenides Foundation International Planetarium Script Competition. In March 1997, *Rusty Rocket's Last Blast* was awarded the First Place Certifi-

The Planetarium also has a Starlab portable planetarium which travels throughout Middle Tennessee on an all-star tour.

Photo: Cumberland Science Museum

cate of Excellence from the Tennessee Association of Museums.

The Sudekum Planetarium generally presents public programs six days a week September through May, and daily, Memorial Day through Labor Day. Planetarium admission is $3 for planetarium only or $1 in addition to regular museum admission of $4.50 for children 3 to 12 and senior citizens, and $6 for 13 and up. The Museum also features two floors of hands-on exhibits, a blockbuster exhibit gallery, and a variety of programs.

Summerhays Planetarium

Brigham Young University
Department of Physics and
Astronomy
Provo, UT 84602-4602
Phone: (801)-378-5396
Internet: physc1.byu.edu/~astrosoc/

The Sarah B. Summerhays Planetarium is located on the fourth floor of the Eyring Science Center (closed for renovation until early 1998) on the Brigham Young University (BYU) campus in Provo, Utah. It is operated by the Department of Physics and Astronomy, with help from the BYU Astronomical Society. Shows are given by faculty and club members.

The observation deck is located on the roof of the science center and is normally only open on Friday nights after the planetarium shows. There is an observatory dome on the very top of the Eyring Science Center. It will contain a 14" reflecting telescope when the renovations are complete.

The BYU Astronomical Society plans to present planetarium shows at 7:30 and 8:30 p.m. on Friday nights. There will be a $2 admission charge per person. On clear Friday evenings, the telescope and observation deck will be open to the general public free of charge. For more information call the planetarium information line.

Taylor Planetarium

Montana State University
Museum of the Rockies
Bozeman, MT 59717
Phone: (406) 994-2251
E-mail:
ammmm@gemini.oscs.montana.edu

The Ruth and Vernon Taylor Planetarium is part of the Museum of the Rockies, a nationally recognized institution dedicated to the physical and cultural interpretation of the Northern Rocky Mountain region. The plan-

etarium opened in April, 1989 as part of a major museum expansion, and has welcomed more than 600,000 visitors to its programs in its first 7 years of operation.

The planetarium theater features a Digistar projection system housed under a 40-foot wide hemispherical projection screen. The theater seats 104 in reclined, upholstered seats in epicentric rows. The Taylor Planetarium presents main feature programs on a variety of astronomy and space-related subjects which change about three times a year. In addition, a live-narrated program called "Rocky Mountain Skies," which explores the current night sky, is presented on weekends and changes seasonally. Laser shows featuring laser images and special effects to rock music are offered on a periodic basis. Children's programs for lower-elementary aged children and their families are offered on Saturday mornings. The planetarium also offers a series of school programs designed for specific grade ranges and topics, available by reservation on weekdays during the school year.

Taylor Planetarium shows are part of a museum astronomy program that includes four STARLAB portable planetariums and two telescope teaching trunks circulated by the museum to schools throughout the state of Montana. Astronomy classes for children and adults, teacher workshops, and special events of astronomical interest are scheduled periodically, and the facility is also used for university astronomy class demonstrations. The planetarium staff also conducts seasonal sky observation sessions for the public using its 8-inch Schmidt-Cassegrain and 13-inch Dobsonian reflecting telescopes.

The planetarium is open daily during the winter season (September to May), with shows on weekday afternoons, regularly on weekends, and on Friday and Saturday nights. School programs may be scheduled on weekdays by reservation. During the summer season (Memorial Day to Labor Day), public performances run regularly every day and evening.

Thomas Jefferson High School for Science and Technology Planetarium

Lee Ann Hennig, Planetarium Teacher
6560 Braddock Road
Alexandria, VA 22312

The Thomas Jefferson High School for Science and Technology Planetarium opened in September of 1967 and features a Spitz A3P instrument and a 30-foot diameter dome with 50 fixed seats in a unidirectional seating arrangement.

Tombaugh Observatory

University of Kansas
Lawrence, KS 66045
Phone: (913) 864-3166/4626
Fax: (913) 864-5262
E-mail: shawl@ukans.edu
Internet: kustar.phsx.ukans.edu:8000/
~shawl/observatory/index.html

The Clyde W. Tombaugh Observatory features a 27-inch Pitt reflector; 6-inch Alvan Clark refractor; 14-inch Celestron; and numerous 8-inch telescopes. Weekly observing sessions are available to the public. Call for days and times.

Trailside Nature & Science Center Planetarium

William P. McClain, Instructor
452 New Providence Rd.
Mountainside, NJ 07092
Phone: (908) 789-3670
Fax: (908) 789-3270

Photo: Trailside Nature & Science Center

The Trailside Nature & Science Center's Planetarium is nestled in the foothills of the 2,000 acre Watchung Reservation. Opened in 1969, it was financed and constructed by the Trailside Museum Association and seats approximately 40 people. The 18-foot diameter dome has recently been upgraded with a Nova Star projector. The planetarium serves the neighboring communities with Sunday family programming at 2 and 3:30 p.m., as well as the monthly "Night Out With the Stars," which includes a topical planetarium show and out-of-doors viewing. The 2 p.m. program changes bimonthly and is limited to ages 6 & up. The 3:30 program alternates between preschool programs for ages 4-6 (with an adult). Admission to these programs is $3 per person. Laser shows for ages 10 and up are also available at $3.25 per person. Senior citizen rates are available and the planetarium is accessible for people with disabilities. School programs can be arranged for Tuesday through Friday, and birthday parties for ages 5 and up can be scheduled by calling (908) 789-3670 between 9 a.m. and 1 p.m., Monday through Friday.

U.S. Naval Observatory

3450 Massachusetts Avenue, NW
Washington, D.C. 20392
Public Affairs Office: (202) 653-1541

Located off Massachusetts Avenue, the U.S. Naval Observatory (USNO) has public nights that feature orientation tours on the USNO's mission, viewing of operating atomic clocks, and glimpses through the finest optical telescopes in the National Capital region.

Source: Art Today

An early photo of the U.S. Naval Observatory's Interior

U.S. Space & Rocket Center

One Tranquility Base
Mail: P.O. Box 070015
Huntsville, AL 35805-3399
Phone: (205) 837-3400
Fax: (205) 837-6137

Photo: U.S. Space & Rocket Center

Take a ride on the Space Shot

A real adventure awaits those planning a visit to the U.S. Space & Rocket Center in Huntsville, Alabama. Start with dozens of hands-on exhibits and more than 1,500 artifacts, followed by *Space Shot*, which allows you to experience more G-force than the astronauts during a shuttle launch as you shoot up a 180-foot tower in only two-and-a-half seconds. Now that you've experienced real G's, how about flying a mission to Mars aboard the simulated space probe "Explorer" which moves as it carries you from a space station to Mars and back again. *Journey to Jupiter* is an action-packed voyage to our largest planet inside a motion-based simulator. You'll be transported to the future and inoculated by lasers before the Helix Catapult zaps you to deep space. If none of this is to your liking then test your aviation skills and see if you can land the Space Shuttle (simulated, of course).

> *dozens of hands-on exhibits and more than 1,500 artifacts*

Other features at the U.S. Space & Rocket Center include *Outpost in Space: The International Space Station*, a theatrical demonstration about living and working in zero-gravity; the Spacedome Theater, featuring IMAX Dome movies; *Shuttle to To-*

Photo: U.S. Space & Rocket Center

Rocket Park

morrow, which takes you on a simulated shuttle mission to a space station; Space Camp and Space Academy, where you'll see future space explorers as they train (see the Space Camp listing under *Products & Ser-*

Awesome place!

vices for more details); an impressive list of Apollo artifacts, including the Lunar Rover, Saturn I rocket (standing 18-stories tall), Saturn V rocket, and the Apollo 16 capsule; the nation's only full-scale Space Shuttle exhibit; the million-gallon Neutral Buoyancy Simulator; three gift shops, and much, much more.

University of Arkansas-Little Rock Planetarium

John Williams, Director
2801 South University Avenue
Little Rock, AR 72204-1000
Phone: (501) 569-3259
Information: (501) 569-3277
Fax: (501) 569-3314
Internet: www.planetarium.ualr.edu/

The University of Arkansas-Little Rock (UALR) Planetarium is the largest planetarium in Arkansas, with a 40-foot dome and reclining seating for 122. The planetarium publishes a quarterly astronomy journal called *dARK Sky*.

University of California - Los Angeles (UCLA) Planetarium

Division of Astronomy
Math Science Building, room 8224
Phone: (310) 825-4434
E-mail: Liles@astro.ucla.edu

The UCLA Planetarium offers three free sky shows a week, Tuesday and Wednesday at 7 p.m. (7:30 p.m. spring and summer) and Sunday at 2 p.m. Seating is limited to 50 per show. A 14-inch telescope is available for public viewing after the sky show every Wednesday. For further information, please contact Laurie Liles via phone or e-mail.

University of Florida Observatory

Department of Astronomy
P.O. Box 112055
Gainesville, FL 32611-2055
Phone: (352) 392-2052 (office)
Fax: (352) 392-5089
E-mail: elton@astro.ufl.edu
Internet: www.astro.ufl.edu

The Department of Astronomy of the University of Florida conducts an open house for the general public on Friday nights between 8:30 and 10:00 p.m., weather permitting and when school is in session. Visitors get a chance to see the Moon, planets, double stars, star clusters, nebulae and other astronomical objects through the 8- and 12-inch telescopes, view images taken with the Hubble Space Telescope, get the latest astronomical news, and ask their own questions. Slide shows on various themes are also presented. This event is free to the public.

The UF Astronomy Teaching Observatory is located south of the Reitz

view images taken with the Hubble Space Telescope

Union parking lot and west of the Aerospace Engineering building, off the Museum Road. Since public viewings are sensitive to weather conditions, please call (352)-392-5294 after 7:30 p.m. for a recorded message.

University of North Alabama Planetarium

Dr. David R. Curott, Director
University of North Alabama
UNA Box 5150
Florence, AL 35632
Phone: (205) 765-4334
Fax: (205) 765-4329
E-mail: dcurott@unanov.una.edu

The University of North Alabama (UNA) Planetarium seats 70. Groups may schedule a presentation (call for current charges, dates, and times).

Valdosta State University Planetarium

Office of University Relations
1500 N. Patterson Street
Valdosta, GA 31698-0215
Phone: (912) 333-5980

Located in Room NH3007 on the third floor of Nevins Hall on the main campus, Valdosta State University

Planetarium seats 65 and is open to the general public. The planetarium's series of evening shows, which begin at 8 p.m. on selected evenings, are free of charge and last about one hour. An observatory open house follows the show, weather permitting. School groups and civic organizations may schedule programs by calling the planetarium at the number provided.

Van Vleck Observatory

Wesleyan University
Middletown, CT 06459
Information: (860) 685-3140

The Van Vleck Observatory operates three telescopes year round for several ongoing research projects,

student instruction, and public observing programs. Their 20 inch refractor (located in the largest dome of the observatory) was supposed to be installed when the current observatory building was dedicated in 1914, but the delivery of the German-made glass blanks for the lenses was delayed by World War I. After almost a decade, the telescope was installed in 1922, and was discovered to have a very high glass quality, allowing a 20" aperture rather than the 18.5" that had been ordered.

On display in the main lobby you'll find the observatory's first telescope, a Fisk Refractor, circa 1836. Now over 160 years old, this telescope was used until its retirement in 1986. Wesleyan University's Astronomy Department offers an undergraduate major program and a graduate Master's degree.

Vega-Bray Observatory

Skywatcher's Inn
Benson, AZ
c/o 420 S. Essex Lane
Tucson, AZ 85711
Phone: (520) 745-2390
Fax: (520) 745-2390
E-mail: vegasky@azstarnet.com
Internet: www.communiverse.com/
skywatcher

Photo: Skywatcher's Inn

Vega-Bray Observatory is located at Skywatcher's Inn

The Vega-Bray Observatory is a privately owned amateur astronomy observatory dedicated to public education. The facilities are modern and house a variety of telescopes ranging from 6-inches to 20-inches in diameter. Some of these are computerized and are able to slew to a galaxy or nebula on command. The observatory is available to the guests of Skywatcher's Inn (see separate listing in the *Products & Services* section).

Vega-Bray Observatory also house a small science museum and a planetarium that are part of the teaching program. Situated at 3800 feet elevation, the observatory was founded in 1990 and is used by various amateur astronomers from southern Arizona. It is also used as a teaching facility for school children of all ages, college students, scouts, social groups and many others.

Situated 47 miles from Tucson, the Inn and Observatory are in an area that offers some of the best dark skies in the USA while being relatively close to the comforts of civilization. While good weather for observing can never be guaranteed, southern Arizona has one of the highest percentages of clear nights in the USA and the world.

Vega-Bray Observatory also house a small science museum and a planetarium that are part of the teaching program

133

Virginia Air & Space Center

Kim L. Maher, Executive Director
600 Settlers Landing Road
Hampton, VA 23669-4033
Phone: 1-800-296-0800
   or (757) 727-0900
Fax: (757) 727-0898
Internet: www.vasc.org

more than 100 historic, aeronautic and space exhibits

From historic air and space craft suspended below the Center's 94-foot ceiling to giant-screen IMAX films, your imagination will soar when you visit the official visitor center for NASA Langley Research Center. Open seven days a week, this world-class facility is located on Hampton's beautiful, downtown waterfront and holds more than 100 historic, aeronautic and space exhibits. NASA artifacts include the Apollo 12 Command Module, a three-billion-year-old moon rock and a Mars meteorite. Approximately one-quarter of a million people visit the Center every year.

Summer hours (Memorial Day - Labor Day) are Monday - Wednesday, 10 - 5; and Thursday - Sunday, 10 - 7. Winter hours are Monday - Sunday, 10 - 5. Exhibit admission is $6 for adults, $4 for seniors, military, and children ages 3 to 11. Exhibit admission plus the IMAX film is $9 for adults, $8 for seniors and military, and $7 for children ages 3 to 11.

Von Braun Planetarium

P.O. Box 1142
Huntsville, AL 35807
Phone: (205) 464-0945
E-mail: adamsml@pipeline.com
Internet: members.aol.com/
VBASTRSOC/VBAS.html
Members: 150

The Wernher von Braun Planetarium was built in the 1960s by the Rocket City Astronomical Association (now known as the Von Braun Astronomical Society—see separate listing under *Organizations*). The planetarium, along with two observatories, is located on a 13.5 acre tract of land in Monte Sano State Park.

The Von Braun Planetarium is used for a variety of public programs, many of which are presented by Von Braun Astronomical Society (VBAS) members. Other speakers are drawn mostly from the Huntsville community (which includes the University of Alabama in Hunstville and NASA's Marshall Space Flight Center), but special guests from outside the area often make presentations. Programs in the planetarium are scheduled at bi-weekly intervals, at 7:30 p.m. Admission is $2.00 for adults, $1.00 for children, 6-12, and members are admitted free of charge. Please contact the Von Braun Astronomical Society for a current schedule, or see their web page.

Wallace Planetarium

Scott Young, Director
1000 John Fitch Highway
Fitchburg, MA 01420
Phone: (508) 343-7900
Fax: (508) 343-8217
E-mail: syoung@net1plus.com/users/
Internet: www.net1plus.com/syoung/
agwp.html

The Alice G. Wallace Planetarium features a 40-foot dome, seating for 103 and a gift shop. The planetarium serves the public as well as the schools in the area. A unique offering of programs (day and overnight) are available for Scout groups. The Wallace Planetarium is part of the Wallace Civic Center Complex which houses two ice-skating arenas. This allows the Wallace Planetarium to offer to school groups Field Trip events which combine a Planetarium show, lunch in the Galaxy Room and skating in one of the arenas. For the current public show schedule or to make reservations for school, scout or private groups call (508) 343-7900.

Wallops Flight Facility Visitor Center
Tours: (757) 824-2298

The NASA Goddard Space Flight Center's Wallops Flight Facility Visitor Center showcases the world of past, present and future flight beginning with ancient Chinese and Greek stories of flight, through the start of powered flight, and into the modern era with the first flight of the Space Shuttle, and beyond. The center is located on Virginia Route 175, about six miles from Route 13, and five miles from Chincoteague, Virginia.

The NASA WFF Visitor Center hosts 60,000 guests of all ages annually. Exhibits include current and future NASA projects, full-scale aircraft and rockets, scale models of space probes, satellites and aircraft, and a moon rock brought back to Earth from the Apollo 17 mission. A number of hands-on exhibits allow you to build and launch a sounder rocket, design your own aircraft, and gain an understanding of the speed of data transmission via satellites.

Tours are available for schools and civic groups and may be arranged throughout the year. Programs can include lectures, films, tours of the complex on Wallops Island, and tours of other areas of the Wallops Flight Facility. Each program is geared toward the age, education level, and interests of the group's members.

The Visitor Center is open from 10 a.m. to 4 p.m. Thursday through Monday (daily during July and August). Admission is free to Visitor Center activities. For further information on all programs, call (757) 824-2298.

A Teacher Resource Laboratory (TRL) is also available which includes publications and videotapes for use by educators (grades K through 12). The TRL is open from 10 a.m. to 4 p.m. on Fridays and Saturdays from June through November. All services are free. Publications cover a variety of topics, including planets, astronomy, aeronautics, the Space Shuttle, life in space, careers, and the environment. Educators wishing to copy videotapes for use in the classroom must provide their own VHS tapes. Because of a limited number of duplication machines, reservations are requested for videotaping and may be made by calling the TRL at (757) 824-1776.

Wells Planetarium

James Madison University
Harrisonburg, VA 22807
Internet: doc.jmu.edu/physics/
planet.htm

The John C. Wells Planetarium is located in Miller Hall on the James Madison University campus. The planetarium has unidirectional seating for 65 under a 30-foot dome. The planetarium is used as a laboratory for JMU's liberal studies and observational astronomy courses. Planetarium programs are available for elementary and secondary public school groups. Contact Rob Grube at 540-JMU-3845 for further information. Public planetarium programs of current interest are presented every Thursday evening at 7 and 8 p.m. There is no admission charge. Call 540-JMU-STAR for further information.

West Ottawa Planetarium

Sandra Plakhe, Director
West Ottawa Middle School
3700 140th Avenue
Holland, MI 49424
Phone: (616) 786-2030

Located at the West Ottawa Middle School, the West Ottawa Planetarium offers programs for area schools, scouts, church groups, and civic organizations. The planetarium is also the gathering place for the Shoreline Amateur Astronomers Association, which meets the third Thursday of each month.

Western Connecticut State University Observatory
Danbury, CT 06810

Construction of the new University Astronomical Observatory at the Westside campus to house a 20-inch Ritchey-Chretien telescope is now complete. The observatory is located on the top of a 5-acre hill with an elevation of 744 feet on what is now called Observatory Hill. The Dome is 16-ft in diameter and 20 feet above the ground with the central pier isolated from the building to avoid telescope vibration. The newly built observatory will greatly strengthen Western's astronomical research and instructional programs, and will offer increased community services. In April 1996, the Perkin Fund awarded the Observatory a grant of $36,000 for the purchase of a large format CCD camera for the 20-inch telescope.

Bi-weekly public viewing nights are conducted by Dr. Dennis Dawson every 1st and 3rd Friday night during semesters. Public nights for special events will be conducted using the automated 20-inch telescope during summer months. Arrangements can be made for night observation and planetarium shows for groups of more than 40 by request.

Western Washington University Planetarium
MS 9064 Physics/Astronomy Dept.
Bellingham, WA 98225
Phone: (360) 650-3818
E-mail: skywise@cc.wwu.edu
Internet: www.wwu.edu/~skywise

The planetarium dome at Western Washington University is 24 feet in diameter and has ample seating for 45 adults. The projector is a Spitz A-2, which projects 1000 naked eye stars, planets, the sun, the moon in 10 different phases, the Milky Way, Andromeda, a comet, and other views of the northern and southern skies.

NOTE: The planetarium is closed for renovations, though it is expected to reopen sometime in 1998. Upon re-opening, presentations can be arranged for groups such as Scouts, clubs, dorms, and classes. Family nights will occur as regularly scheduled events. Shows last about an hour and are free. The planetarium show focuses on constellations, naked eye and binocular astronomy, and lots of Native American starlore. Call for information on group reservations or to reserve space for family night.

Westminster College Planetarium

Hoyt Science Resources Center
Physics Department
Westminster College
New Wilmington, PA 16172
Phone: (412) 946-7200
Fax: (412) 946-7146
E-mail: lightner@westminster.edu

The Westminster College Planetarium is associated with the Physics Department at Westminster College. It has a 24-foot dome and can accommodate groups of up to 50. The planetarium supports the college's astronomy course, while providing educational opportunities for other classes. The planetarium also does presentations for area school groups and offers public shows (admission is free). Call for information or to be placed on the mailing list.

West Springfield High School Planetarium

John Dieringer, Planetarium Teacher
6100 Rolling Road
Springfield, VA 22151
Phone: (703) 913-3889

The West Springfield High School Planetarium was installed in March of 1967 and features a Spitz A3P projector and 67 seats in a chevron seating pattern.

Wheaton College Observatory

Timothy Barker, Professor of Astron.
Wheaton College
Norton, MA 02766
Phone: (508) 286-3975
Recorded Message: (508) 286-3937
Fax: (508) 285-8278

The Wheaton College Observatory is open to the public every clear Friday night that the College is in session. Located on the roof of the Science Center at Wheaton, the Observatory has seven computerized 12-inch telescopes and two computerized 14-inch telescopes. Call for a recorded message with current information.

Woodson Planetarium

Horizons Unlimited
1636 Parkview Circle
Salisbury, NC 28144
Phone: (704) 639-3004
Fax: (704) 639-3015
E-mail:
Patsy.Wilson@mail.rowan.k12.nc.us

The Margaret C. Woodson Planetarium seats 76 people under a 30-foot dome and features a Spitz A3P star projector. The shows offered range from local live star shows to multi-media productions. The planetarium is part of Horizons Unlimited, a 19,000 square foot hands-on museum that serves as a supplementary education center in the Rowan-Salisbury school system. The center offers innovative learning opportunities for children, youths and adults.

In addition to the Woodson Planetarium, other space-related features include NASA's Apollo Lunar Lander Display and the Roof-Top Observatory. Programs offered at the center include special events, teacher workshops, summer activities for children, classes for special groups (from preschoolers to senior citizens), and school programs.

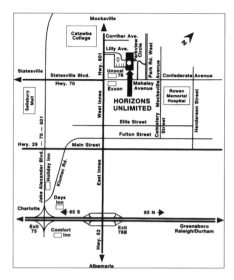

Other Places of Interest

There were a number of astronomy and space-related sites from which we did not receive information. Though we can't be certain the information presented here is correct, we've provided the names and addresses of these sites as a service to our readers.

Akima Planetarium
East Tennessee Discovery Center
P.O. Box 6204
Knoxville, TN 37914

Alden Planetarium
New England Science Center
222 Harrington Way
Worcester, MA 01604-1899
Internet: www.nesc.org

Aldrin Planetarium
South Florida Science Museum
4801 Dreher Trail North
West Palm Beach, FL 33405

Allegheny College Planetarium
Physics Department
Meadville, PA 16335

Allen Memorial Planetarium
Vigo County School Corporation
3737 South 7th Street
Terre Haute, IN 47802

Alma College Planetarium
Dow Science Building
Alma, MI 48801

Alworth Planetarium
University of Minnesota - Duluth
10 University Drive
Duluth, MN 55812

Anderson Planetarium
Lambuth University
705 Lambuth Boulevard
Jackson, TN 38301

Andrews Planetarium
Deerfield Academy
Deerfield, MA 01342

Andrus Planetarium
Hudson River Museum
511 Warburton Avenue
Yonkers, NY 10701
Phone: (914) 963-4550
Planetarium Hotline: (914) 963-2139

Angelo State University Planetarium
Department of Physics
San Angelo, TX 76909

Apache Point Observatory
2001 Apache Point
Sunspot, NM 88349

Armstrong Planetarium
Altoona Area High School
6th Avenue & 15th Street
Altoon, PA 16602

Astronaut Memorial Planetarium & Observatory
Brevard Community College
1519 Clearlake Road
Cocoa, FL 32922

Austin State University Planetarium
Department of Physics and Astronomy
P.O. Box 13044
Stephen F. Austin Station
Nacogdoches, TX 75962

Ball State University Planetarium
Department of Physics and Astronomy
Muncie, IN 47306

Bays Mountain Center & Planetarium
853 Bays Mountain Park Road
Kingsport, TN 37660

Beecher Planetarium
Youngtown State University
Youngstown, OH 44555

Blake Planetarium
117 Long Pond Road
Plymouth, MA 02360-2660

Bonisteel Observatory & Planetarium
Dickenson College
S. C. Box 156
Carlisle, PA 17013

Bowling Green State University Planetarium
Physics and Astronomy Department
Bowling Green, OH 43403

Brazosport Nature Center Planetarium
400 College Drive
Lake Jackson, TX 77566-3146
Internet: www.tgn.net/-snark/ncap/ncap.html

Burke Planetarium
Burke High School
12200 Burke Street
Omaha, NE 68154

Calusa Nature Center Planetarium
3450 Ortiz Avenue
Mail:
P.O. Box 06023
Fort Myers, FL 33911-6023

Carmel High School Planetarium
520 East Main Street
Carmel, IN 46032

Carr-Fles Planetarium
Muskegon Community College
221 South Quarterline Road
Muskegon, MI 49442

Carver High School Planetarium
60 South Meadow Road
Carver, MA 02330

Catawba Science Center
243 Third Avenue N.E.
Hickory, NC 28603

Central Bucks School District Planetarium
Central Bucks East High School
Holicong and Anderson Roads
Buckingham, PA 18912-9999

Central Dauphin High School Planetarium
4600 Locust Lane
Harrisburg, PA 17109-4498

Central Texas College Planetarium
P.O. Box 1800
Killeen, TX 76540

Chadron State College Planetarium
Math and Science Building
Chadron, NE 69337

Ching Planetarium
Hartnell Community College
156 Homestead Avenue
Salinas, CA 93901

Colonial Middle School Planetarium
716 Belvoir Road
Norristown, PA 19401-2577

Copernicus Planetarium
Lourdes Colleges
6832 Convent Boulevard
Sylvania, OH 43560

Cormack Planetarium
Museum of Natural History
Michael Umbricht, Coordinator
Roger Williams Park
Providence, RI 02905
Phone: (401) 785-9457
Fax: (401) 461-5146
E-mail: Cormack_PL@ids.net
Internet: ids.net/-cormack_pl/
museum.html

Cosmic Theater
Richland College
12800 Abrams Road
Dallas, TX 75023

143

County College of Morris Planetarium
214 Centre Grove Road
Cohen Hall
Randolph, NJ 07869

Dale Planetarium
Wayne State College
1111 Main Street
Wayne, NE 68787

Dearborn Observatory
Northwestern University
2131 Sheridan Road
Evanston, IL 60208-2900
Phone: (847) 491-3685
Fax: (847) 491-9982
Internet: www.astro.nwu.edu

Dibert Planetarium
Franklin Science Center
Shippensburg University
Shippensburg, PA 17257

Diehl Planetarium
Route 11
Danville, PA 17821-9811

The Discovery Museum Planetarium
Sacramento Science Center
3615 Auburn Boulevard
Sacramento, CA 95821

Don Harrington Discovery Center
1200 Streit Drive
Amarillo, TX 79106

Dupont Planetarium
Ruth Patrick Science Center
University of South Carolina
171 University Parkway
Aiken, SC 29801

East Penn Planetarium
Emmaus Junior High School
660 Macungie Avenue
Emmaus, PA 18049-2130

Eastern College Planetarium
1300 Eagle Road
Saint Davids, PA 190887-3936

Erie Historical Museum and Planetarium
356 West Sixth Street
Erie, PA 16507

Elgin Observatory & Planetarium
School District U-46
355 East Chicago Street
Elgin, IL 60120

Erie Planetarium
Erie Historical Museum
356 West 6th Street
Erie, PA 16507-1245

Ethyl Universe Planetarium
Science Museum of Virginia
2500 West Broad Street
Richmond, VA 23220
Internet:smv.mus.va.us/
wunihome.html

Exhibit Museum Planetarium
University of Michigan
1109 Geddes Avenue, Room 4506
Ann Arbor, MI 48109-1079
Internet: www.exhibits.lsa.umich.edu

Faulkner Planetarium
Herrett Center for Arts & Science
College of Southern Idaho
315 Falls Avenue
Twin Falls, ID 83303

**Fayetteville State University
Planetarium**
1200 Murchison Road
Newbold Station
Fayetteville, NC 28301

Fick Observatory
1128 240th Street
Boone, IA 50036

Flynn Planetarium
Davis and Elkins College
100 Campus Drive
Elkins, WV 26241

**Fort Lupton High School
Planetarium**
530 Reynolds Street
Fort Lupton, CO 80621

**Francis Marion University
Planetarium**
Hwy 301 North
Florence, SC 29501

**Franklin and Marshall College
Planetarium**
North Museum
Lancaster, PA 17604

Freedom Planetarium
Freedom High School
3149 Chester Avenue
Bethlehem, PA 18017-2896

**Frostburg State University
Planetarium**
Department of Astronomy
Frostburg, MD 21532

Ft. Couch School Planetarium
515 Fort Couch Road
Upper Saint Clair, PA 15241-2018

Fulton Planetarium
2025 Jonesboro Road S.E.
Atlanta, GA

Gardner Memorial Planetairum
Kingman Museum of Natural History
Battle Creek, MI 49017

Gates Planetarium
Museum of Natural History
2001 Colorado Boulevard
Denver, CO 80205

Gengras Planetarium
Science Museum of Connecticut
950 Trout Brook Drive
West Hartford, CT 06119-1437

Georgetown College Planetarium
400 East College Street
Georgetown, KY 40324

Glendale High School Planetarium
1466 Beaver Valley Road
Flinton, PA 16640-8900

Goddard Planetarium
Roswell Museum & Art Center
11th & North Main
Roswell, NM 88201

Golden Pond Planetarium
100 Van Morgan Drive
Land Between the Lakes
Golden Pond, KY 42211

Good Hope Planetarium
Good Hope Middle School
451 Skyport Road
Mechanicsburg, PA 17055-6898

Grady Planetarium
Lehigh Valley Amateur
Astronomical Society
620A East Rock Road
Allentown, PA 18103-7525

**Grand Haven Junior High
School Planetarium**
1400 South Griffin Street
Grand Haven, MI 49417

Hallstrom Planetarium
Indian River Community College
3209 Virginia Avenue
Fort Pierce, FL 34981

Harper Planetarium
Atlanta Public Schools
3399 Collier Drive, N.W.
Atlanta, GA 30331

Hefferan Planetarium
Albuquerque Public Schools
Career Enrichment Center
807 Mountain Road, N.E.
Albuquerque, NM 87102-2441

**Hooper Planetarium &
Daniel Observatory**
Roper Mountain Science Center
504 Ropper Mountain Road
Greenville, SC 29615

Horwitz Planetarium
School District of Waukesha
222 Maple Avenue
Waukesha, WI 53186

Howell Memorial Planetarium
Bob Jones University
1700 Wade Hampton Boulevard
Greenville, SC 29614

Hudnall Planetarium
Tyler Junior College
P.O. Box 9020
Tyler, TX 75711-9020
Internet: www.tyler.cc.tx.us/Showcase/
Tour/Planet/planet.htm

Hurst Planetarium
Jackson High School
544 Wildwood Avenue
Jackson, MI 49201

Illinois State University Planetarium
Physics Department
College Avenue & School Street
Normal, IL 61790-4560

**Independence High School
Planetarium**
1776 Educational Park Drive
San Jose, CA 95133

International Space Hall of Fame
Tombaugh Space Theater
Top of Highway 2001
Alamogordo, NM 88310
Phone: (505) 437-2840
Internet: abcz.nmsu.edu/-bwood/

Jones Planetarium & Observatory
University of Tennessee
at Chattanooga
10 North Tuxedo Avenue
Chattanooga, TN 37403

Keck Observatory
65-1120 Mamalahoa Highway
Kamuela, HI 96743

Keeble Observatory
Department of Physics
Randolph-Macon College
P.O. Box 5005
Ashland, VA 23005-5505
Phone: (804) 752-7344
Fax: (804) 752-4724
E-mail: gspagna@rmc.edu
Internet: www.rmc.edu/-gspagna/
gspagna.html

Kent County High School Planetarium
25301 Lambs Meadow Road
Worton, MD 21678

Kent State University Planetarium
Physics Department
Kent, OH 44242

King Planetarium
King Science Center at Mann
3720 Florence Boulevard
Omaha, NE 68110

Kirk Planetarium
State University College at New Paltz
New Paltz, NY 12561

Kountze Planetarium
University of Nebraska at Omaha
Physics Department
6010 Dodge Street
Omaha, NE 68182-0266
Internet: www.unomaha.edu/-stlucas/index.html

Layfayette Natural History Museum Planetarium
637 Girard Park Drive
Layfayette, LA 70503

Lane Education Service District Planetarium
2300 Leo Harris Parkway
Eugene, OR 97401-8834

Lansing Community College Planetarium
419 North Capital Avenue
Lansing, MI 48910

Liberty Science Center
251 Phillip Street
Jersey City, NJ 07304

Los Medanos College Planetarium
2700 East Leland Road
Pittsburg, CA 94565-5197

Lueninghoener Planetarium
Midland Lutheran College
900 North Clarkson
Fremont, NE 68025

Madison Metro School District Planetarium
James Madison Memorial High School
201 South Gammon Road
Madison, WI 53717-1499
Internet: www.madison.k12.wi.us/planetarium/

Malcolm Science Center Planetarium
P.O. Box 186
Easton, ME 04740

Mallon Planetarium
Arcola Intermediate School
1 Eagleville Road
Norristown, PA 19403-1477

Mattingly Planetarium
Hiahleah-Miami Lakes
Senior High School
7977 West 12th Avenue
Hialeah, FL 33014-3595

McDonald Planetarium
Hastings Museum
1330 North Burlington Avenue
Hastings, NE 68901

McDonald Planetarium
SER McDonald School
666 Reeves Lane
Warminster, PA 18974-3025

**Milham Planetarium &
Hopkins Observatory**
Williams College
Williamstown, MA 01267

Montour High School Planetarium
Clever Road
McKeese Rocks, PA 15136

Moody Planetarium
Museum of Texas Tech University
Fourth and Indiana
Lubbock, TX 79409

**Mt. Lebanon High School
Planetarium**
155 Cochran Road
Pittsburgh, PA 15228-1381

Mt. Wilson Observatory
740 Holladay Road
Pasadena, CA 91106

Muncie Schools Planetarium
Central High School
801 North Walnut Street
Muncie, IN 47305-1458

Murdock Sky Theater
Oregon Museum of
Science & Industry
1945 S.E. Water Avenue
Portland, OR 97214-3356

**Museum of Arts & Science
Planetarium**
1040 Museum Boulevard
Daytona Beach, FL 32114

National Solar Observatory
P.O. Box 62
Sunspot, NM 88349
Phone: (505) 434-7000

Navigator Planetarium
Fountain of Youth Museum
155 Magnolia Avenue
Saint Augustine, FL 32084

Neil Armstrong Museum
500 South Apollo Drive
Wopakoneta, OH 45895

Neil Armstrong Planetarium
Altoona Area School District
1415 6th Avenue
Altoona, PA 16602

**Newburgh Free Academy
Planetarium**
201 Fullerton Avenue
Newburgh, NY 12550

Newhard Planetarium
The University of Findlay
1000 North Main Street
Findlay, OH 45840

**New Jersey State Museum
Planetarium**
205 West State Street
Trenton, NJ 08625

North Atlanta Planetarium
2875 Northside Drive N.W.
Atlanta, GA 30305-2807

**Northampton Area School District
Planetarium**
1617 Laubach Avenue
Northampton, PA 18067-1517

Northern Stars Planetarium
4 Osborne Street
Fairfield, ME 04937

Ohio State University Planetarium
174 West 18th Avenue
Columbus, OH 43210

**Ohio's Center of Science &
Industry (COSI)**
280 East Broad Street
Columbus, OH 43215

Olson Planetarium
University of Wisconsin - Milwaukee
1900 E. Kenwood
Mail:
P.O. Box 413
Milwaukee, WI 53201-0413
Show Times: (414) 229-4961
Internet: miller.cs.uwm.edu/uwm.html

Palomar College Planetarium
1140 West Mission Road
San Marcos, CA 92069-1487

Palomar Observatory
35899 Canfield Road
Palomar Mountain, CA 92060

Parrott Planetarium
Dolores S. Parrott Middle School
19220 Youth Drive
Brooksville, FL 34601-8600

Patterson Planetarium
Muscogee County School District
2900 Woodruff Farm Road
Columbus, GA 31907

**Philips Space Theater &
Apollo Observatory**
Dayton Museum of Natural History
2600 DeWeese Parkway
Dayton, OH 45414

Phillips Planetarium
Department of Physics and Astronomy
University of Wisconsin
Eau Claire, WI 54702

Pike High School Planetarium
6701 Zionsville Road
Indianapolis, IN 46268

**Pine-Richland High School
Planetarium**
700 Warrendale Road
Gibsonia, PA 15044

**Pittsburgh State University-
Greenbush Astrophysical
Observatory**
P.O. Box 189
947 W 57 Hwy
Girard, KS 66743
Phone: (316) 724-6281
Fax: (316) 724-6284/8475
E-mail: laura.boone@greenbush.org
Internet: www.greenbush.org

**San Francisco State University
Planetarium**
1600 Holloway Avenue
San Francisco, CA 94132-1722

**Reading School District
Planetarium**
1211 Parkside Drive South
Reading, PA 19611-1441

Ricks College Planetarium
Physics Department
Rexburg, ID 83460

Ridley School District Planetarium
1001 Morton Avenue
Folsom, PA 19033-2997

**Ritter Planetarium & Observatory,
Brooks Observatory**
University of Toledo
2801 West Bancroft Street
Toledo, OH 43606
Recording: (419) 530-4037

Riverview High School Planetarium
One Ram Way
Sarasota, FL

Robinson Planetarium
Adrian College
Peelle Hall, Williams Street
Adrian, MI 49221-2575

Rockford Planetarium
Discovery Center Museum
711 North Main Street
Rockford, IL 61103

Rocky Mount Planetarium
Children's Museum
1610 Gay Street
Rocky Mount, NC 27804

Saint Charles Parish Library Planetarium
105 Lakewood Drive
Luling, LA 70070

Sale Planetarium
Virginia Military Institute
Physics & Astronomy Department
Lexington, VA 24450

Sandy Run Middle School Planetarium
520 Twining Road
Dresher, PA 19025-1995

Sanford Museum Planetarium
117 East Willow Street
Cherokee, IA 51012

San Juan College Planetarium
4601 College Boulevard
Farmington, NM 87402

Saunders Planetarium
Museum of Science & Industry
4801 East Fowler Avenue
Tampa, FL 33617

Savannah Science Museum Planetarium
4405 Paulsen Street
Savannah, GA 31405

Schaefer Planetarium
Bettendorf High School
3333 18th Street
Bettendorf, IA 52722

Schiele Planetarium
Museum of Natural History
1500 East Garrison Boulevard
Gastonia, NC 28054

Schreder Planetarium
Shasta County Office of Education
1644 Magnolia Avenue
Redding, CA 96001

The Science Place Planetarium
1620 First Avenue
Mail:
P.O. Box 151469
Dallas, TX 75315
Internet: www.scienceplace.org/planetarium.htm

SciWorks Planetarium
Science Center & Environmental Park of Forsyth County
400 Hanes Mill Road
Winston-Salem, NM 27105

Scobee Planetarium
1300 San Pedro Avenue
San Antonio, TX 78212

Settlemyre Planetarium
Museum of York County
4621 Mt. Gallant Road
Rock Hill, SC 29732

Seymour Planetarium
Springfield Science Museum
236 State Street
Springfield, MA 01103-1778

Sharpe Planetarium
Memphis Pink Palace Museum
3050 Central Avenue
Memphis, TN 38111

Shiras Planetarium
Marquette Senior High School
1203 West Fair Avenue
Marquette, MI 49855

Southern Conneticut State University Planetarium
501 Crescent Street
New Haven, CT 06515-1330

Space & Science Theater
Pensacola Junior College
1000 College Boulevard
Pensacola, FL 32504-8998

State Museum of Pennsylvania Planetarium
3rd & North Street
Mail:
P.O. Box 1026
Harrisburg, PA 17108-1026

Storer Planetarium
Calvert County Public Schools
600 Dares Beach Road
Prince Frederick, MD 20678

Sunrise Museum Planetarium
746 Myrtle Road
Charleston, WV 25314

Tessman Planetarium
Rancho Santiago Community College
17th & Bristol Street
Santa Ana, CA 92706

Tipton Planetarium
Tipton Community School Corp.
817 South Main Street
Tipton, IN 46072-9775

Tomchin Planetarium
West Virginia University
220 Hodges Hall
Morgantown, WV 26505
Internet: www.as.wvu.edu/-planet

Turkey Run State Park Planetarium
Department of Natural Resources
Turkey Run State Park, RR 1
Marshall, IN 47859

West Allegheny Planetarium
West Allegheny Middle School
408 West Allegheny Road
Imperial, PA 15126

Ulmer Planetarium
Lock Haven University of Pennsylvania
Fairview Street
Lock Haven, PA 17745

West Chester University Planetarium
Geology and Astronomy Department
West Chester, PA 19383

University of Wisconsin Planetarium - La Crosse
Cowley Hall
La Crosse, WI 54601

Westminster College Planetarium
Department of Physics
New Wilmington, PA 16172-0001

Wetherbee Planetarium
100 Roosevelt Avenue
Albany, GA 31701-2325

Victor Valley College Planetarium
18422 Bear Valley Road
Victorville, CA 92392-5849
Wagner College Planetarium
Grymes Hill
631 Howard Avenue
Staten Island, NY 10301

Wichita Falls Museum & Art Center
2 Eureka Circle
Wichita Falls, TX 76308

Walker County Science Center Planetarium
322 Highway 95
Rock Spring, GA 30739

Wichita Omnisphere & Science Center
220 South Main Street
Wichita, KS 67202

Washington County Planetarium & Space Science Center
823 Commonwealth Avenue
Hagerstown, MD 21740

Wilderness Center Planetarium
9877 Alabama Avenue, S.W.
Wilmot, OH 44689

Wiley Planetarium
Delta State University
P.O. Box 3255
Cleveland, MS 38733

**York Suburban High School
Planetarium**
Hollywood Drive and Southern Road
York, PA 17403-3097

Yerkes Observatory
373 W. Geneva Street
Williams Bay, WI 53191

Zacheis Planetarium & Observatory
Adams State College
208 Edgemont Boulevard
Alamosa, CO 81101

Zane Planetarium
Natural Science Center
4301 Lawndale Drive
Greensboro, NC 27455

Places of Interest
QuickFind Index
A state by state listing

Alabama
Gayle Planetarium
Marshall Space Flight Center
U.S. Space & Rocket Center
University of North Alabama Planetarium
Von Braun Planetarium

Arizona
Discovery Park, Gov Aker Observatory
Flandrau Science Center Planetarium
  and Observatory
Lowell Observatory
Pomeroy Planetarium
Vega-Bray Observatory

Arkansas
University of Arkansas - Little Rock
  Planetarium

California
Ames Research Center
Bakersfield College Planetarium
California State University, Northridge
  Planetarium
Chabot Observatory & Science Center
Ching Planetarium
Discovery Museum Planetarium
Dryden Flight Research Center
Fremont Peak Observatory
Gladwin Planetarium
Griffith Observatory

Holt Planetarium
Independence High School Planetarium
Jet Propulsion Laboratory
Los Medanos College Planetarium
Minolta Planetarium
Morrison Planetarium
Mount Wilson Observatory
Palmer Observatory
Palomar College Planetarium
Palomar Observatory
San Francisco State University
  Planetarium
Reuben H. Fleet Space Theater &
  Science Center
Rosicrucian Egyptian Museum and
  Planetarium
San Diego Aerospace Museum
Santa Rosa Junior College
  Planetarium
Schreder Planetarium
Tessman Planetarium
University of California - Los Angeles
  (UCLA) Planetarium
Victor Valley College Planetarium

Colorado
Air Force Academy Planetarium
Fiske Planetarium
Fort Lupton High School Planetarium
Gates Planetarium
Johnson Planetarium
Sommers-Bausch Observatory
Zacheis Planetarium and Observatory

Connecticut
Foran High School Planetarium
Gengras Planetarium
Southern Connecticut State University
  Planetarium
Van Vleck Observatory
Western Connecticut State University
  Observatory

Patterson Planetarium
Rollins Planetarium
Savannah Science Museum
  Planetarium
Smith Planetarium
Valdosta State University Planetarium
Walker County Science Center
  Planetarium
Wetherbee Planetarium

Delaware
Starwalk Planetarium

Hawaii
Onizuka Space Center
Keck Observatory

Florida
Air Force Space and Missile Museum
Aldrin Planetarium
Astronaut Memorial Planetarium &
  Observatory
Bishop Planetarium
Buehler Planetarium
Calusa Nature Center Planetarium
Hallstrom Planetarium
Kennedy Space Center
Mattingly Planetarium
Museum of Arts and Science
  Planetarium
Navigator Planetarium
Parrott Planetarium
Riverview High School Planetarium
Robinson Observatory
Saunders Planetarium
Space and Science Theater
University of Florida Observatory

Idaho
Faulkner Planetarium
Ricks College Planetarium

Illinois
Adler Planetarium & Astronomy
  Museum
Cernan Earth and Space Center
Dearborn Observatory
Elgin Observatory and Planetarium
Illinois State University Planetarium
Lakeview Museum of Arts and
  Sciences
Rockford Planetarium
Staerkel Planetarium
Strickler Planetarium

Georgia
Fernbank Science Center
Fulton Planetarium
Harper Planetarium
North Atlanta Planetarium

Indiana
Allen Memorial Planetarium
Ball State University Planetarium
Carmel High School Planetarium
Holcomb Observatory and
  Planetarium
Hook Astronomical Observatory

Kirkwood Observatory
Koch Science Center and Planetarium
Merrillville Community Planetarium
Muncie Schools Planetarium
Pike High School Planetarium
SpaceQuest Planetarium
Tipton Planetarium
Turkey Run State Park Planetarium

Iowa
Gale Observatory
Sanford Museum Planetarium
Sargent Space Theater
Schaefer Planetarium
Fick Observatory

Kansas
Kansas Cosmosphere and
  Space Center
Kelce Planetarium
Pittsburg State University-Greenbush
  Astrophysical Observatory
Tombaugh Observatory
Wichita Omnisphere and Science
  Center

Kentucky
Georgetown College Planetarium
Golden Pond Planetarium
Hummel Planetarium
Rauch Memorial Planetarium

Louisiana
Challenger Learning Center
Freeman Astronomy Center
Freeport McMoRan Daily Living
  Science Center Planetarium
  and Observatory

Lafayette Natural History Museum
  Planetarium
Saint Charles Parish Library
  Planetarium

Maine
Jordan Planetarium and Observatory
Malcolm Science Center Planetarium
Northern Stars Planetarium
Southworth Planetarium

Maryland
Davis Planetarium
Frostburg State University Planetarium
Goddard Space Flight Center
Kent County High School Planetarium
McKeldin Memorial Planetarium
Montgomery College Planetarium
Owens Science Center
Storer Planetarium
Washington County Planetarium and
  Space Science Center

Massachuesetts
Aldrin Planetarium
Andrews Planetarium
Blake Planetarium
Carver High School Planetarium
Connecticut River Valley Astronomers'
  Conjunction
Harvard-Smithsonian Center for
  Astrophysics
Hayden Planetarium
Milham Planetarium and Hopkins
  Observatory
Seymour Planetarium
Wallace Planetarium
Wheaton College Observatory

Michigan

Abrams Planetarium
Alma College Planetarium
Carr-Fles Planetarium
Chaffee Planetarium and
  Veen Observatory
Exhibit Museum Planetarium
Gardner Memorial Planetarium
Grand Haven Junior High School
  Planetarium
Hurst Planetarium
Jesse Besser Museum Planetarium
Lansing Community College
  Planetarium
Longway Planetarium
Michigan Space and Science Center
Robinson Planetarium
Sherzer Observatory
Shiras Planetarium
West Ottawa Planetarium

Minnesota

Alworth Planetarium
College of St. Catherine Observatory
Minneapolis Planetarium
Paulucci Space Theater
Saint Cloud State University
  Observatory and Planetarium

Mississippi

Davis Planetarium
Kennon Observatory
Stennis Space Center
Wiley Planetarium

Missouri

Challenger Learning Center
McDonnell Planetarium

Montana

Taylor Planetarium

Nebraska

Burke Planetarium
Chadron State College Planetarium
Dale Planetarium
King Planetarium
Kountze Planetarium
Lueninghoener Planetarium
McDonald Planetarium
Mueller Planetarium
Seven Hills Observatory

Nevada

Community College of Southern
  Nevada Planetarium
Fleischmann Planetarium

New Hampshire

Christa McAuliffe Planetarium

New Jersey

County College of Morris Planetarium
Dreyfuss Planetarium
Glenfield Planetarium
Liberty Science Center
New Jersey State Museum
  Planetarium
Novins Planetarium
Raritan Valley Community College
  Planetarium
Stockton College Observatory
Trailside Nature & Science Center
  Planetarium

New Mexico

Apache Point Observatory
Goddard Planetarium
Hefferan Planetarium
International Space Hall of Fame
National Solar Observatory
Planetarium at Santa Fe Community
  College
San Juan College Planetarium

New York

Andrus Planetarium
Custer Institute
Kirk Planetarium
Link Planetarium
Longwood Regional Planetarium
Newburgh Free Academy Planetarium
New York Hall of Science
Schenectady Museum and
  Planetarium
Silverman Planetarium
Stull Observatory
Wagner College Planetarium

North Carolina

Catawba Science Center
Fayetteville State University
  Planetarium
Kelly Space Voyager Planetarium
Rocky Mount Planetarium
Schiele Planetarium
SciWorks Planetarium
Woodson Planetarium
Zane Planetarium

Ohio

Beecher Planetarium
Bowling Green State University
  Planetarium
Copernicus Planetarium
Kent State University Planetarium
Neil Armstrong Museum
Newhard Planetarium
Ohio State University Planetarium
Ohio's Center of Science & Industry
Perkins Observatory
Philips Space Theater and
  Apollo Observatory
Ritter Planetarium and Observatory,
  Brooks Observatory
Wilderness Center Planetarium

Oklahoma

Kirkpatrick Planetarium
Oklahoma Baptist University
  Planetarium

Oregon

Chemeketa Community College
  Planetarium
Lane Education Service District
  Planetarium
Murdock Sky Theater
Pine Mountain Observatory

Pennsylvania

Allegheny College Planetarium
Allentown School District Planetarium
Armstrong Planetarium
Bonisteel Observatory and
  Planetarium
Buhl Planetarium and Observatory
Central Bucks School District
  Planetarium
Central Dauphin High School
  Planetarium
Colonial Middle School Planetarium

Detwiler Planetarium
Dibert Planetarium
Diehl Planetarium
East Penn Planetarium
Eastern College Planetarium
Edinboro University Planetarium
Erie Historical Museum and
  Planetarium
Erie Planetarium
Fels Planetarium
Franklin and Marshall College
  Planetarium
Freedom Planetarium
Ft. Couch School Planetarium
Glendale High School Planetarium
Good Hope Planetarium
Grady Planetarium
Indiana University of Pennsylvania
  Planetarium
Mallon Planetarium
McDonald Planetarium
Montour High School Planetarium
Mt. Lebanon High School Planetarium
Museum of Scientific Discovery
Neil Armstrong Planetarium
Northampton Area School District
  Planetarium
Pine-Richland High School
  Planetarium
Reading School District Planetarium
Ridley School District Planetarium
Sandy Run Middle School Planetarium
State Museum of Pennsylvania
  Planetarium
Ulmer Planetarium
West Allegheny Planetarium
West Chester University Planetarium
Westminster College Planetarium
York Suburban High School
  Planetarium

Rhode Island

Cormack Planetarium

South Carolina

Dupont Planetarium
Francis Marion University Planetarium
Gibbes Planetarium
Hooper Planetarium and
  Daniel Observatory
Howell Memorial Planetarium
Settlemyre Planetarium
Stanback Planetarium

Tennessee

Akima Planetarium
Anderson Planetarium
Bays Mountain Nature Center and
  Planetarium
Jones Planetarium and Observatory
Sharpe Planetarium
Sudekum Planetarium

Texas

Angelo State University Planetarium
Austin State University Planetarium
Brazosport Nature Center Planetarium
Burke Baker Planetarium
Central Texas College Planetarium
Cosmic Theater
Don Harrington Discovery Center
El Paso ISD Planetarium
Hudnall Planetarium
Johnson Space Center
Moody Planetarium
Noble Planetarium
Science Place Planetarium
Scobee Planetarium
Space Center Houston
Wichita Falls Museum and Art Center

Utah

Hansen Planetarium
Ott Planetarium
Summerhays Planetarium

Virginia

Brackbill Planetarium
Carl Sandburg Middle School
  Planetarium
Chesapeake Planetarium
Ethyl Universe Planetarium
Falls Church High School Planetarium
Hayfield Secondary School
  Planetarium
Herndon High School Planetarium
Keeble Observatory
Oakton High School Planetarium
Peninsula Planetarium and
  Abbitt Observatory
Sale Planetarium
Thomas Jefferson High School for
  Science and Technology Planetarium
Virginia Air & Space Center
Wallops Flight Facility Visitor Center
Wells Planetarium
West Springfield High School
  Planetarium

Washington

Goldendale Observatory
Smith Planetarium
Western Washington University
  Planetarium

Washington, D.C.

Einstein Planetarium
NASA Headquarters
National Air and Space Museum
Rock Creek Park Planetarium
U.S. Naval Observatory

West Virginia

Berkeley County Planetarium
Flynn Planetarium
Sunrise Museum Planetarium
Tomchin Planetarium

Wisconsin

Barlow Planetarium
Horwitz Planetarium
Madison Metro School District
  Planetarium
Olson Planetarium
Phillips Planetarium
University of Wisconsin Planetarium -
  La Crosse
Yerkes Observatory

Wyoming

Casper Planetarium

Organizations

An alphabetical listing of astronomy clubs, space exploration societies, and research organizations. For a state by state listing, refer to the QuickFind Index that follows this section.

2111 Foundation for Exploration

P.O. Box 338
Mountain View, CA 94042-0338
Phone: 1-888-THE-2111
E-mail: Foundation@twentyone-11.org
Internet: www.twentyone-11.org

From remote sensing of forests to studying humans in extreme environments to prepare for the exploration of Mars. From studying biospheres in space to satellite imaging of glaciers to study global warming. The connections between Space Exploration and Earth Preservation are both wide and deep and critical for our progress into the 21st century.

The Twenty-one Eleven Foundation is a non-profit organization dedicated to forging these connections. It publishes a biannual magazine called *Tranquility*, the Magazine of Environmental and Space Exploration. This publication is available free to members and is written in an engaging and non-technical style covering a whole range of articles linked to Earth and Space Exploration. Each year the Foundation awards ten '2111 Awards' to projects around the world that have demonstrated outstanding contributions to forging the balance between Space and Environmental Exploration. The Foundation also launches K-12 education programs linked to its work.

Dues: $25 annual

Abilene Astronomical Society

1817 Jackson Street
Abilene, TX 79602-4635
Phone: (915) 677-9781
Members: 15

Organized on the campus of Hardin-Simmons University in 1955, the Abilene Astronomical Society continues to meet at the university and offers membership benefits that include a discount rate on *Sky & Telescope* magazine and use of club telescopes.

Dues: $2 annual

The Albuquerque Astronomical Society

P.O. Box 54072
Albuquerque, NM 87153-4072
Hotline: (505) 296-0549
Bulletin Board Service: (505) 867-4295 (8N1)
Internet:
 www.phys.unm.edu/~egates/TAAS/taas.html
Members: 300+

Members of The Albuquerque Astronomical Society (TAAS) include both amateur and professional astronomers and the Society has programs for all interest levels. Meetings are held monthly at the University of New Mexico in Regener Hall. TAAS owns and operates the General Nathan Twining Observatory (GNTO) at a dark sky site near Belen, New Mexico. General Nathan Twining donated the land for the facility and construction took place in 1991 - 1992. The observatory now houses the Isengard telescope (a 16-inch Cave Newtonian) donated by Lt. Col. Bill Isengard. Star parties are held at the observatory at least monthly and the site is available for TAAS member use at any time. Special events held at the General Nathan Twining Observatory include Ladies Night and the Messier Marathon.

Membership benefits include a subscription to the *Sidereal Times*, the Society's monthly newsletter, access to club telescopes, check-out privileges from the extensive TAAS library, and more. An Amateur Telescope Making Class is held twice yearly (reduced fees for members).

Dues: $20 annual (regular membership)

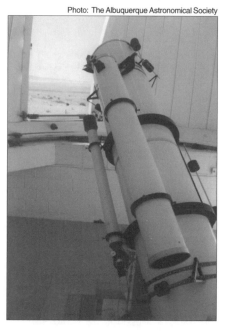

Photo: The Albuquerque Astronomical Society

Inside the General Nathan Twining Observatory (GNTO)

Amarillo Astronomy Club

Robert Ashcraft
7913 Mitcham Drive
Amarillo, TX 79121
E-mail: rashcraf@arn.net
Members: 35

The club has a dark observing site about one hour from Amarillo

Founded in 1981, the Amarillo Astronomy Club holds meetings once a month between September and May at the Discovery Center in Amarillo's Medical Complex, usually near the full moon. Planned monthly observing sessions are held as are informal observing sessions. The club has a dark observing site about one hour from Amarillo that has a support building which houses the club's library and equipment. The site also has a computer with *MegaStar*, *The Sky*, and *The Real Sky*, plus other programs. Star parties are held twice a year at Caprock Canyons (for a full weekend of amateur astronomy). Member benefits include the club's newsletter, *Amarillo Astronomy Chronicles*, discount subscription rates on *Sky & Telescope* and *Astronomy* magazines, use of club telescopes, access to the observing site, and membership in the Astronomical League (through club affiliation).

Amateur Astronomers, Inc.

William Miller Sperry Observatory
1033 Springfield Ave.
Cranford, NJ 07016
Phone: (908) 709-7520
Contact: George Chaplenko
(908) 549-0615
Information (taped): (908) 276-STAR
Internet: idt.net/~lewycky/aai.html
Members: 300

AAI operates the William Miller Sperry Observatory, a facility that features a members-built 10-inch refractor telescope and a 24-inch Cassegrain telescope.

Founded on November 14, 1949, Amateur Astronomers, Inc., (AAI) grew from a small group of enthusiasts to one of the largest organizations of its kind, with close to 300 members. AAI's main purpose is promoting interest in the science of astronomy through a variety of activities and educational programs. These include regularly-scheduled Adult Education courses; monthly Members meetings (open to public) featuring invited speakers; and an E.T. Pearson college scholarship. In addition, a number of Star Parties are held throughout the year, including one during the annual Astronomy Day event. AAI is widely known for its successful Solar Eclipse Expeditions.

AAI operates the William Miller Sperry Observatory, a facility that features a members-built 10-inch refractor telescope and a 24-inch Cassegrain telescope. The club also publishes the *Qualified Observer's Handbook* (for members who participate in the annual Qualified Observer's Course); *Asterism*, a newsletter, published nine times a year; and *Sperry Observations*, an aperiodic scholarly journal covering the scientific work of members.

The club is organized into a number of interest and service Committees, supporting all areas of astronomy, education and public relations. Special group discounts are available on subscriptions to *Astronomy* and *Sky & Telescope* magazine.

Dues: $17 annual (regular)
         $27 annual (sustaining)
         $42 annual (sponsor)

Amateur Observers' Society of New York, Inc.

Susan Rose, President
P.O. Box 500
Merrick, NY 11566-0500
Phone: (516) 489-2970
Internet: members.aol.com/
rjbenjamin/aos.html
Members: 73

The Amateur Observers' Society of New York, Inc. (AOS) was established in 1965 and has active members from ages 7 to 80. Meetings are held on the first Sunday of the month (October through May), and the first Saturday of June and September. At least two observing sessions are held each month and a picnic is scheduled for the Perseid Meteor Shower.

The AOS is a member organization of the Astronomical League and is very active in bringing astronomy to the general public. The Hall of Science of The City of New York has requested their participation in a program sponsored by the Astronomical Society of the Pacific, whereby they instruct educators on how to teach astronomy, and provide observing sessions for students. Members receive the *Celestial Observer*, the Society's monthly newsletter.

Dues: $15 annual (primary member)
$5 annual (additional member)
$10 annual (associate member)

Amateur Telescope Makers Association

17606 28th Avenue S.E.
Bothell, WA 98012
Phone: (206) 481-7627
Fax: (206) 281-4921
E-mail: atmj1@aol.com

This association is for those who want to build their own telescope for a fraction of what it would cost to purchase a comparable unit. The organization publishes the *Amateur Telescope Making Journal*, which includes articles on the latest optical and mechanical innovations.

Dues: $22 annual (U.S.)
$30 annual (Canada)
$32 annual (overseas)

American Astronomical Society

2000 Florida Avenue, N.W., Suite 400
Washington, D.C. 20009
Phone: (202) 328-2010
Fax: (202) 234-2560
E-mail: aas.aas.org
Internet: www.aas.org
Members: 6,500

The American Astronomical Society is the major organization of professional astronomers in the United States, Canada, and Mexico. The basic objective of the AAS is to promote the advancement of astronomy and closely related branches of science. The membership, now at approximately 6,500, also includes physicists, mathematicians, geologists, engineers and others whose research interests lie within the broad spectrum of subject matter now comprising contemporary astronomy.

Amherst Area Amateur Astronomers Association

1403 South East Street
Amherst, MA 01002-3031
Phone: (413) 256-6234
Fax: (413) 256-6234
E-mail: astro@crocker.com
Internet: www.tiac.net/users/ whitney/5a
Members: 70

Mt. Pollux Conservation Area, the Amherst Area Amateur Astronomers Association, or 5As, is an active and involved astronomy association. The club sponsors a number of events and programs, including Astronomy Day, public outreach and school outreach programs, and a variety of observatory and planetarium programs. 5As facilities and all their programs are open to the public and are free. Meetings are held the second and fourth Friday of each month at 8 p.m.

With facilities that include the Bassett Planetarium and Wilder Observatory at Amherst College, and the

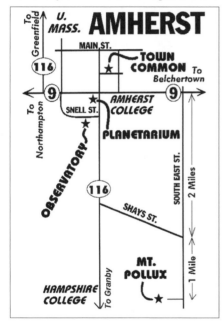

How active is 5As? Well, programs sponsored by the association attract 12,000 to 14,000 attendees each year, they publish their own observing guides, write regular newspaper columns, and do radio and television programs.

Membership benefits include free observing guides, software, a computer link, charts, maps, a calendar, and a subscription to the association newsletter. Members also receive discounts on magazine subscriptions, books, software, telescopes, and equipment.

Dues: $10 annual

Ancient City Astronomy Club

P.O. Box 546
St. Augustine, FL 32085-0546
Phone: (904) 446-4338
Members: 63

Founded in 1974, the Ancient City Astronomy Club meets the first Friday of each month at 8 p.m. at the St. Au-

gustine High School Observatory. A member of the Astronomical League, the club established a "Herschel Club" and in 1980 published a manual, *Guide to Herschel Objects*, which is now available as an official publication of the Astronomical League. The club offers free public observing sessions on a regular basis and gives free "Introduction to Astronomy" classes for schools and interested groups. Membership benefits include access to a 12.5" reflecting telescope, a 6" reflector, a 60-mm and two 75-mm refractors, an Edmund 4 1/4" Astroscan 2001, a 16" Meade LX200 (at the St. Augustine High School Observatory), and a subscription to the clubs bimonthly newsletter, *Cosmic Echoes*.

Dues: $12 annual (individual)
$15 annual (family)
$7 annual (student, age 12-18)

Association of Universities for Research in Astronomy

Suite 550
1625 Massachusetts Ave. N.W.
Washington, D.C. 20036
Members: 25 U.S. universities and 3 International Affiliates

The Association of Universities for Research in Astronomy (AURA) is a consortium of educational and other non-profit institutions which operate world-class astronomical observatories. Founded in 1957, the association endeavors to serve the entire astronomical community by providing modern, major facilities to qualified scientists from all universities. To this end, they operate three centers: the National Science Foundation's (NSF) National Optical Astronomy Observatories (NOAO); NASA and the European Space Agency's Space Telescope Science Institute (STScI), which operates the Hubble Space Telescope (HST); and the International Gemini Project.

Membership in AURA is open to all U.S. academic institutions and non-profit organizations that meet the guidelines. International Affiliates were added as a new category of membership in 1992 for institutions outside the U.S. If your institution is interested in membership, contact Lorraine Reams, Director of Corporate Relations, for information.

Association of Lunar and Planetary Observers - Meteors Section

161 Vance Street
Chula Vista, CA 91910-4828
E-mail: LUNRO.IMO.USA@prodigy.com
Internet: www.lpl.arizona.edu/alpo/

The Meteors Section of the Association of Lunar and Planetary Observers (ALPO) was reorganized in 1989 and now aligns itself with the International Meteor Organization (IMO) and the North American Meteor Network. The goals of the Meteor Section are to promote meteor observing and to assist in the training of observers new to the field. Data received from observers are published in the quarterly ALPO journal, *The Strolling Astronomer*, and then forwarded to the IMO for further analysis.

The Meteor Section publishes a pamphlet, *The A.L.P.O. Guide for Observing Meteors*, which is available from the Astronomical League Sales (P.O. Box 572, West Burlington, IA 52655) for $3 plus $1 postage. The pamphlet contains the basic information needed to conduct a meteor watch plus observing forms and a current calendar to help plan your activities.

The Association of Lunar and Planetary Observers is divided into sections representing the major interest areas of the solar system. There are coordinators for each of the planets, the sun, the moon, comets, and meteors.

Dues: $16 annual

Astronomical League

Janet Stevens, Executive Secretary
2112 Kingfisher Lane East
Rolling Meadows, IL 60008
Mail:
Science Service Building
1719 N. Street, N.W.
Washington, DC 20036
Phone: (847) 398-0562
Fax: (847) 607-8636 (call first)
E-mail: jastevens@compuserve.com
Internet: www.mcs.net/~bstevens/
al/index.html
Members: 215 organizations
     13,500 individuals

Founded in 1947 by Dr. Harlow Shapley, the Astronomical League (AL) is a federation of astronomical societies with the stated purpose of promoting the science of astronomy by fostering astronomical education, by providing incentives for astronomical observation and research, and by assisting communication among amateur astronomical societies. Most of the 13,000 members of AL are conferred membership through their local astronomical society.

Benefits of membership in the Astronomical League include a subscription to the *Reflector*, the AL's quarterly newsletter, access to the AL's Book Service whereby members can obtain books and publications at 10% off list price (with no shipping and handling charges), and the opportunity to take part in eclipse trips and an annual star party in the southern hemisphere. The Astronomical League also publishes *Observe* manuals which are guides to observing the sky, instructional *AstroNotes*, and other materials useful to members, clubs and public educators.

AL's quarterly publication

The Astronomical League sponsors a number of Observing Clubs which offer certificates of accomplishment for demonstrating observing skills. These Observing Clubs include the Messier Club, The Herschel 400 Club, Binocular Messier Club, Deep Sky Binocular Club, the Southern Skies Binocular Club, the Sunspotter Club, the Meteor Club, and the Lunar Club. In addition, the National Young Astronomer Award (NYAA) was established in 1991 for high school students aged 14-19. The winner of this award receives a 10" Meade LX200 telescope, a trip to the Astronomical League national convention, and a lifetime observing pass to share telescope time with professionals at the McDonald Observatory in Texas.

The easiest way to join the Astronomical League is by joining a member society in your area. For a list of Astronomical League member societies, see Appendix A. Memberships-at-large are also available for $10 (annual) and include full voting privileges, a subscription to the *Reflector*, Book Service access, magazine discounts, and the opportunity to earn observing awards.

The Astronomical Society of Australia

ASA Secretary
School of Physics,
University of Sydney
NSW 2006
Australia
Phone: 61-2-9351-3184
Fax: 61-2-9351-7727
E-mail: j.obyrne@physics.usyd.edu.au
**Internet: www.atnf.csiro.au/
asawww/asa/html**
Members: 350

Formed in 1966, the Astronomical Society of Australia (ASA) is the body of professional astronomers in Australia. Members are mostly active professional astronomers and postgraduate students; however, some retired astronomers and distinguished amateur astronomers are also members. The *Publication of the Astronomical Society of Australia* (PASA) is a peer-reviewed journal published as one volume of three issues per year.

Astronomical Society of Kansas City

P.O. Box 400
Blue Springs, MO 64013
StarTouch Line: (889-STAR, #5400)
Reservation Line: (913) 438-3825
E-mail: askc@sound.net
Internet: www.sound.net/~askc/
Starry Night BBS: (913) 631-0761

ASKC operates two observatories; the Powell Observatory and the Elmcrest Observatory

The Astronomical Society of Kansas City (ASKC) began as the Amateur Astronomers and Telescope Makers of Kansas City back in 1935. The Society is a member of the Astronomical League and holds meetings on the fourth Saturday of each month (except December), at 7 p.m. in room 103 of Royal Hall, on the University of Missouri, Kansas City campus. Those wishing to attend should call the StarTouch Astronomy line to confirm the meeting time and place.

The ASKC operates two observatories; the Powell Observatory and the Elmcrest Observatory. The Powell Observatory is located near Louisburg, Kansas and features a 30-inch computer controlled f/5 Newtonian on a fork equatorial mount. A heated classroom with restrooms is attached to the 20-foot domed observatory. Public programs are presented each month, May through October, usually on the first and third Saturdays. The observatory can also be reserved for private programs, or star parties by private clubs or groups.

The Elmcrest Observatory is ASKC's second observatory. It was completed in 1992 on the grounds of the Powell Botanical Gardens with the help of the Powell Family Foundation. The observatory was dismantled in 1996 to make way for a new visitor's center and plans are underway to relocate the facility to a new location in the Gardens. Even with the temporary absences of the Elmcrest Observatory, ASKC members continue to present "Star Light, Star Bright" public programs at the Gardens on the third Friday of each month at the Trigg Building. Reservations are required and can be made by calling (816) 697-2600, extension 225.

The Astronomical Society of Kansas City publishes the *Cosmic Messenger*, their monthly newsletter, and members also receive the quarterly Astronomical League publication, the *Reflector*. Other membership benefits include access to Powell Observatory for personal observing, access to books, tapes and slides from the club library, discounts on astronomy and other science-related books, and discounts on *Astronomy* and *Sky & Telescope* magazines.

Dues: $25 annual (individual)
       $30 annual (family)

Astronomical Society of Las Cruces

Roy Willoughby, President
P.O. Box 921
Las Cruces, NM 88004
Phone: (505)523-7370 (Treasurer)
E-mail: willowbee@zianet.com
Internet: www.zianet.com/silvern/aslc/aslcp1.htm
Members: 80

The Astronomical Society of Las Cruces New Mexico (ASLC) was formed in 1951 and has grown to approximately 80 members. Founders of the Society include Jed Durrenberger; Walter Haas, founder of the Association of Lunar and Planetary Observers (ALPO) and long time Editor of the ALPO newsletter, *The Strolling Astronomer*, and Dr. Clyde Tombaugh, discoverer of the planet Pluto. Jed Durrenberger and Walter Haas are still Directors Emeritus of ASLC. Dr. Clyde Tombaugh passed away January 17, 1997.

This Society was formed to promote amateur astronomy and provide a means of inspiring newcomers to get involved in this fascinating and exciting hobby. The Society provides and sponsors informative and educational programs geared to the amateur, many provided by expert professional astronomers from the Astronomy Department of New Mexico State University and from other parts of the country.

Meetings are usually scheduled for the third or fourth Friday of each month. Star parties are usually scheduled for sunset on the Saturday preceding the New Moon. Everyone interested in astronomy is invited to attend these meetings and star parties. You don't need a telescope or binoculars or even knowledge of how to use them. Many telescopes are usually available to look through, and the owners are always anxious to show you what they are looking at and share their knowledge.

Dues: $12 (Membership only)
       $39 (Membership + subscription to *Sky & Telescope*)

Astronomical Society of Long Island, Inc.

Ron Rizzi, President
13 Lincoln Avenue
Massapequa, NY 11758
Phone: (516) 795-3072
Members: 150

Since 1964, the Astronomical Society of Long Island (ASLI) has drawn members from Long Island and adjacent areas, bringing together people from all walks of life. Prospective members do not need to know anything about astronomy, nor do they need a telescope. Meetings are held every Wednesday at 8:30 p.m. on the campus of the New York Institute of Technology (NYIT), in the Schure Hall Lecture Room. NYIT is located on Northern Boulevard in Old Westbury, New York. Membership includes a one year subscription to *Sky & Telescope* or *Astronomy* magazine, a subscription to *ASLI News*, and *The Observer's Handbook*.

Dues: $45 (individual)
       $20 (student)

Astronomical Society of Nevada

Jack Parker, President
78 Middleton Way
Fernley, NV 89408
Phone: (702) 575-4889
E-mail: j@parker.reno.nv.us
Members: 65

The purpose of the Astronomical Society of Nevada (ASN) is to foster the public's interest in astronomy and science. The Society was founded in 1934 and publishes the *ASN Bulletin* for members.

The ASN holds a club star party on the Saturday night closest to the new moon. Details on the location of the party may be obtained by calling (702) 323-2404.

Dues: $39 annual (includes subscription to *Sky & Telescope*)

Astronomical Society of Northern New England

P.O. Box 497a
Kennebunkport, ME 04046
Phone: (207) 967-5945
E-mail: Astronomy@nlis.net
Internet: www.nlis.net /~mesky/
Members: 65+

The Astronomical Society of Northern New England (ASNNE) was established in 1982 and meets the first Friday of each month. Star parties are typically held on the new moon weekend each month (note: please call ahead to confirm location and date). Club assets include a 20-inch Dobsonian reflector, a 10-inch Dobsonian reflector, a 16-inch portable equatorial, an equatorial mount for 20-inch, and various other items, including a nice library. Plans are being made to build a permanent observatory at Days Meadow Science Center in West Kennebunk, Maine.

Dues: $20 annual (individual)
$24 annual (family)
$10 annual (student)

The Astronomical Society of the Pacific

390 Ashton Avenue
San Francisco, CA 94112
Phone: (415) 337-1100
Fax: (415) 337-5205
Internet: www.aspsky.org
Members: 6000

Founded over a hundred years ago and with members in all fifty states and seventy countries, The Astronomical Society of the Pacific (ASP) is one of the largest, finest astronomical associations in the world. The ASP is responsible for educational tools such as the quarterly newsletter *The Universe in the Classroom,* which is delivered free of charge to over 10,000 educators worldwide. The Society has several publications, including *Society Scope,* the ASP newsletter, and *Mercury,* the ASP's information-packed bimonthly journal, which includes monthly Sky Calendars and Star Maps for use in the Northern Hemisphere. This magazine is included with membership. Other Society benefits include a 10% discount on all purchases

One of the Society's publications, *Universe in the Classroom,* is distributed to over 10,000 educators worldwide.

from the ASP's impressive catalog, which includes books, posters, videotapes, software, CD-ROMs, slide sets, gift items, and observing aids. Additional benefits are available to professional members (technical/graduate student).

Dues: $35 annual (regular)
$44 annual (foreign)
$80 annual (U.S. technical)
$40 annual (U.S. graduate)

The Astronomical Society of the Toms River Area (ASTRA)

c/o Novins Planetarium
Ocean County College
P.O. Box 2001
Toms River, NJ 08754-2001
Phone: (732) 255-0343
E-mail:
zimmermann@monmouth.com
Members: 85

The Astronomical Society of the Toms River Area (ASTRA) meets the second Friday of each month (except August) at 8 p.m. at the Novins Planetarium. Society telescopes include a 6-inch Dobsonian, an 8-inch Dobsonian, and a 4-inch Astro-Scan. New members pay a one-time assessment of $5 toward the telescope fund. Member privileges include a monthly newsletter, *Astral Projections*, student admission to Planetarium shows, membership in the Astronomical League, and use of club telescopes.

Dues: $10 annual

The Astronomical Unit

P.O. Box 3702
Santa Barbara, CA 93130
Phone: (805) 682-4711 x316
E-mail: kary@io.ucsb.edu

Sponsored by the Santa Barbara Museum of Natural History, The Astronomical Unit welcomes new members of all ages and backgrounds, from avid telescope-users to folks who know very little about the sky, but are curious and want to learn more. The club meets at 7:30 p.m. on the first Friday of each month (with occasional exceptions during the summer) in the Museum's Farrand Hall (2559 Puesta del Sol Road). Club activities and benefits include dark-site observations, public observations and community programs, field trips, planetarium shows, and a newsletter.

Dues: $15 annual (individual)
$25 annual (family)

Astronomy Day Headquarters

**Gary Tomlinson,
Coordinator, Astronomical League
Public Museum of Grand Rapids
272 Pearl NW
Grand Rapids, MI 49504
Phone: (616) 456-3532
Fax: (616) 456-3873
E-mail: gtomlins@triton.net
Internet: www.mcs.net/~bstevens/al**

Astronomy Day Headquarters (operated by the Astronomical League) supplies information and assistance to organizations and groups who desire to host special events for Astronomy Day. This day (held sometime mid-April to mid-May) is set aside each year to celebrate and encourage the science of astronomy. The theme of Astronomy Day is: Bringing Astronomy to the People. Astronomy Day is an international event serving thousands of people at hundreds of locations and is cosponsored by 15 national and international organizations. The Headquarters also administers the *Sky & Telescope* Astronomy Day Award, given each year to the group who provides the best Astronomy Day event. For the exact date, Astronomy Day locations, tips for hosting and access to the 60-page handbook of suggestions for hosting an Astronomy Day event, visit the Astronomy Day Headquarters' web site.

Auburn Astronomical Society

**c/o Mr. John Zachry,
Secretary/Treasurer
P.O. Box 778
West Point, GA 31833-0778**

The regular meeting of the Auburn Astronomical Society is on the first Friday of each month, in room 215, of the Aerospace Engineering building, on the main campus of Auburn University at 8 p.m. Dark-sky observing is scheduled for the weekend nearest the new moon of each month. Membership benefits include discount subscription rates for *Astronomy* and *Sky & Telescope* magazines, access to a safe dark-sky location for observing, newsletters with news of upcoming events, and more. Because the Auburn Astronomical Society is affiliated with the Astronomical League members are entitled to enjoy all of the benefits afforded to League members, including quarterly issues of the *Reflector*.

Dues: $15 annual (regular)
        $ 7.50 annual (student)

Back Bay Amateur Astronomers

2808 Flag Road
Chesapeake, VA 23323-2102
Phone: (757) 485-4242
Fax: (757) 485-4242
E-mail:
GlendonHowell@compuserve.com
Members: 23 member families

Founded in December 1978, the goal of the Back Bay Amateur Astronomers is to promote the science of astronomy as a hobby both for members and for the general public. Primarily an observational group, they hold regular meetings on the first Thursday of each month at various sites. A monthly Skywatch observing program is held on the Friday or Saturday nearest to the New Moon at Northwest River Park in Chesapeake, a reasonably dark site. The group offers its services to Planetariums, Schools, Churches, and Civic Leagues free of charge. They also sponsor a regional star party, the *East Coast Star Party*, typically in the Fall of each year. The organization publishes a monthly newsletter, the *Back Bay Observer*, edited by Glendon L. Howell.

Dues: $18 annual

Battle Point Astronomical Association

P.O. Box 10914
Bainbridge Island, WA 98110
Facility Director: (206) 842-4001
Chairman: (206) 842-3717
E-mail:
egardine@linknet.kitsap.lib.wa.us
Internet: home.sprynet.com/
spryn~/aurorae
Members: 200

Battle Point Astronomical Association (BPAA) is a unique amateur organization dedicated to astronomical education for all ages. Its purpose is to create awareness among students by providing instruction in astronomy on all levels, providing viewing through first class-equipment, supporting school programs both at the Society's observatory and through on-site lectures, providing demonstrations and conferences on astronomy, providing work areas for optical and machine shop equipment, housing astronomical equipment including telescopes and electronic devices, sponsoring Astronomy Day, monthly star parties and full-moon viewing sessions, and setting up an Internet link to other organizations for the transfer of images, data and papers.

BPAA is focusing on hands-on astronomy and education for the young. Telescopes and other devices are designed and built to be used as easily by school children as by adults. The intent is to make astronomy available to everyone, regardless of age or skills. All equipment is built with the end-user in mind.

BPAA has restored/remodeled the Helix House into the Edwin E. Ritchie Observatory and Community Learning Center at Battle Point Park on Bainbridge Island, Washington. The Society is nearing completion of their telescope, built from a world-class zerodur mirror (27.5") donated by Boeing. BPAA members have ground and figured the mirror to astronomical tolerances on grinding and polishing equipment designed and built by Ed Ritchie, their late Chief Astronomer. This equipment will be made available to members for their own construction projects. First light is anticipated in September 1997.

The Society's web site is regularly updated with scheduled events and current club information, plus there's a Kid's Page for younger enthusiasts. Membership benefits include a bi-monthly newsletter, membership in the Astronomical League, free observing information, a computer link, peripheral information, and discounts to sponsored events. Board meetings are held the first Wednesday of each month at 7 p.m. Star Parties are on the Saturday closest to the new moon and Full Moon Fridays are on the Friday closest to the full moon at 8 p.m.

Dues: $15 annual

Birmingham Astronomical Society

1341 Stonehedge Drive
Brimingham, AL 35235
Internet: www.secis.com/
~braeunig/bas.html

The Birmingham Astronomical Society (BAS) provides a program meeting on the third Wednesday of each month at 7 p.m. in the Hoover Public Library that is open to the general public (except December, when the club holds its Christmas Party) and features a video or lecture on an astronomical topic that will both entertain and educate. The BAS supports local educators by sending speakers to local grade schools and by providing astronomy labs for local college classes. While colleges and universities are charged a reasonable fee for astronomy labs, there is no fee for grade school speakers. The club maintains a dark-sky observing site on the crest of Chandler Mountain and publishes a monthly newsletter, *Newscope*, which is mailed to all members.

Dues: $20 annual (individual)
$25 annual (family)
$12 annual (student)

British National Space Centre (BNSC)

151 Buckingham Palace Road
London SW1W 9SS
Phone: 0171 215 0901
Fax: 0171 215 0936
E-mail: Information@BNSC-
HQ.ccMail.CompuServe.com
Internet: www.open.gov.uk/bnsc/
bnschome.htm

Campbell County High School Astronomy Club

Mark L. Brackin, club sponsor
1000 Camel Drive
Gillette, WY 82716
E-mail:
mbrackin@cchs.ccsd.k12.wy.us
Members: 35

The Campbell County High School Astronomy Club is composed entirely of high school students and has the twofold purpose of educating the public through star parties for the community and providing a way of getting telescopes into the hands of students who otherwise could not afford them. There is no cost for joining. The club welcomes donations of any type of telescope or binocular for student use. Such donations are tax deductible.

Catawba Valley Astronomy Club

5026 Poplar View Lane
Granite Falls, NC 28630
LMO Telephone (704-428-9955)

The Catawba Valley Astronomy Club (CVAC) is open to anyone with an interest in astronomy. The club meets at 7:30 p.m. on the second Thursday of each month at either the Lucile Miller Observatory (LMO) on the campus of Maiden High School or in the Executive Dining Room on the ground floor of the Student Services building on the campus of Catawba Valley Community College (CVCC). The club publishes a newsletter, *After Dark*, for member news and information on upcoming events.

Cedar Amateur Astronomers, Inc.

Contact: Doug Slauson
73 Summit Avenue
Swisher, IA 52338
E-mail: dslauson@cedar-rapids.net
Internet: www.geocities.com/
CapeCanaveral/8866/CAA-htm.
Members: 65

The Cedar Amateur Astronomers hosts public nights at the Palisades-Dows Observatory once each month, March through November. An extra observing night is scheduled each month for club members and interested guests.

The Palisades-Dows Observatory is located approximately 10 miles southeast of Cedar Rapids. It has a 12-1/2 inch classical Cassegrain telescope permanently mounted in a 16-foot dome. The observatory is equipped with a CCD camera and the telescope can accept photographic and photoelectric equipment. Other telescopes at the observatory include a 5-inch refractor, a 6-inch refractor, a 12-1/2-inch Newtonian reflector, and a 17-1/2-inch Newtonian reflector. A 18'x24' roll-off roof building is expected to be completed in 1997 to house some of these instruments.

Membership benefits include the monthly newsletter, *The Prime Focus*, a membership packet, access to the observatory, a well-stocked astronomical library, and membership in the Astronomical League.

Dues: $20 annual (individual)
$15 annual (student)
$25 annual (family)

Center for the Study of Extraterrestrial Intelligence

Dr. Steven Greer,
International Director
P.O. Box 15401
Asheville, NC 28813
Fax: (704) 274 6766
Phone: (704) 274 5671
E-mail: webmaster@cseti.org
Internet: www.cseti.org
Members: 3,000

Founded in 1991, the Center for the Study of Extraterrestrial Intelligence (CSETI) is the only worldwide organization dedicated to establishing peaceful and sustainable relations with extraterrestrial life forms.

Write, call, or e-mail for details about membership, programs, or other CSETI facts.

Central Iowa Astronomers

c/o Joanne L. Hailey
1116 42nd Street
Des Moines, IA 50311
Phone: (515) 277-2739

The Central Iowa Astronomers (CIA) was formed in July of 1994 as a means to provide education and promote interest in astronomy. The club meets at 7:30 p.m. on the first Thursday of each month at the Des Moines Science Center. As part of its public outreach, the club provides viewing sessions and lectures for Scouts, schools, church groups, and individuals. The CIA is also affiliated with the Astronomical League.

Membership benefits include a subscription to the *Central Iowa Astronomers Informer*, the club's monthly newsletter, a subscription to the *Reflector*, the quarterly publication of the Astronomical League, and discounts on books and magazines.

Dues: $10 annual (17 and under)
$18 annual (18 and over)
$24 annual (family)
$10 annual (Member-at-Large)

Central Wyoming Astronomical Society

c/o Casper Planetarium
904 Poplar
Casper, WY 82601
Phone: (307) 577-0310
E-mail: cwas@coffey.com
Members: 8

The Central Wyoming Astronomical Society meets the second Tuesday each month at the Casper Planetarium during the school year and at alternative sites during the summer. Though still small, the club is active in providing the community with a source of information on a variety of astronomical subjects, including the proper setup and use of telescopes, observing sessions, and public access for astronomical events.

Champaign-Urbana Astronomical Society and Observatory

c/o Champaign Park District
706 Kenwood Road
Champaign, IL 61821
Phone: (217) 351-2567
E-mail: dleake@parkland.cc.il.us
Internet: www.prairienet.org/
business/astro/cuas.html
Members: 70

Founded in 1986, the Champaign-Urbana Astronomical Society (CUAS) has members of all ages and backgrounds. The group meets the second Thursday of the month at 7 p.m. at the Staerkel Planetarium at Parkland College. The CUAS has a club library and a monthly newsletter, *Clear Skies*, available to members. The club also operates a rural observatory about 15 miles southwest of Champaign. A 15-foot, twin shutter dome houses a home-made 16-inch, f/15 Cassegrain system with a 6-inch refracting telescope piggybacked on top. This refurbished observatory dome was built in 1913 by the University of Illinois and was instrumental in the development of photoelectric photometry, an important astronomical invention pioneered in Illinois. This dome was one of four that focused starlight on a photocell to trip a relay and turn on the lights of the Century of Progress Exposition as part of the Chicago World's Fair in 1933. Open houses are conducted, weather permitting, on the Saturday closest to the first quarter moon, free of charge.

Dues: $15 annual (family)

Cheyenne Astronomical Society

3539 Luther Place
Cheyenne, WY 82001
Phone: (307) 635-5944
E-mail: mcurran@sisna.com
Internet: www.sisna.com/users/
mcurran
Members: 100

Founded in 1986, the Cheyenne Astronomical Society (CAS) holds regular meetings and publishes a 10-page monthly newsletter called the *Cosmic Babbler*. The Society also sponsors an annual "Weekend Under the Stars" at Foxpark, Wyoming.

Dues: $12 annual

Cincinnati Astronomical Society

5274 Zion Road
Cleves, OH 45002
Phone: (513) 941-1981
Members: 180

Founded in 1911, the Cincinnati Astronomical Society (CAS) is a large, active organization that promotes astronomy to the general public through star parties, lectures, astronomical displays, and classes. The Society owns two sites; the first is the 26-acre site where their headquarters building is located (on Zion Road), and the second site is an 18-acre dark sky site in Adams County, Ohio.

The CAS meets at 8 p.m. on the third Friday of each month at the Zion United Methodist Church at Zion and Zion Hill Roads. Informal star gazing at the Zion Road headquarters usually follows the general meetings, when skies are clear.

The CAS headquarters building has complete darkroom facilities, a well-stocked astronomical library, a photo gallery, and computer. The site also features four separate observatory buildings with permanently mounted telescopes. The 18-acre dark sky site in Adams County, Ohio features a camping area, charcoal grill, electricity and The Bob Schultz Memorial Observatory which has two permanently mounted scopes and several other portable scopes available. The CAS owns a nice selection of telescopes ranging in size from 6-inch to 27.5-inch aperture. The Society is affiliated with the Astronomical League and members receive the *Reflector*, the League's quarterly publication.

Dues: $25 annual (individual)
        $30 annual (family)
        $15 annual (18 and under)

Colorado Springs Astronomical Society

P. O. Box 62022
Colorado Springs, CO 80962
Phone: (719) 573-1173
Public Event Line: (719) 444-2567,
message 610
E-mail: frazierm@usa.net
        or bygrens@aol.com
Internet: www.uccs.edu/~cwetheri/
Astro/CSAS/
Members: 70

Each summer, CSAS hosts the Rocky Mountain Star Stare, an astro camp-out in the Tarryall Range west of Colorado Springs.

The Colorado Springs Astronomical Society (CSAS) serves as a gathering point for people interested in all aspects of astronomy. In addition to club star parties, CSAS hosts public viewing nights during important astronomical events. These sessions are held at city parks or sports stadiums and draw thousands of spectators and media coverage. Each summer, CSAS hosts the Rocky Mountain Star Stare, an astro camp-out in the Tarryall Range west of Colorado Springs. This event provides excellent dark sky observing and draws attendees from around the world. CSAS members work together with the physics department of the University of Colorado at Colorado Springs (UCCS) to help maintain the university observatory and present Astronomy Day events. Members also work in cooperation with state parks, community recreational groups, and schools to present educational astronomy sessions.

Colorado Springs Astronomical Society members receive monthly issues of the club's newsletter, *The Hypoxic Observer*, membership in the Astronomical League and a subscription to the *Reflector*, the AL newsletter. CSAS members also receive discounts on astronomical products and magazine subscriptions. In addition, members have the opportunity to earn viewing awards and certificates through the Astronomical League.

Club meetings are held at 7 p.m. on the fourth Tuesday of each month and are open to the public. Meetings consist of a business session followed by a presentation or guest speaker. Elections of club officers are held once a year. CSAS hosts club star parties once a month, plus numerous spontaneous viewing sessions organized through the club's E-mail list and event line. For more information on CSAS, access their web site at www.uccs.edu/~cwetheri/Astro/CSAS/.

Dues: $17 annual

Columbus Astronomical Society

P.O. Box 449
Delaware, OH 43015
Phone: (614) 363-1257

The Columbus Astronomical Society normally meets at 8 p.m. on the second Saturday of each month at the Perkins Observatory. The group is active in the community and sponsors some of the programs at Perkins Observatory. For a copy of the latest CAS newsletter and membership information, send a stamped, self-addressed, business-sized envelope.

Comox Valley Astronomy Club

Site 514-1758 Greenwood Cres.
RR5 Comox, B.C. V9N 8B5
Canada

Founded in 1990, the primary purpose of the Comox Valley Astronomy Club is visual viewing, though some members are active in astro photography. The club has two reflector telescopes, an 8 inch f7 reflector with 1¼ inch focuser and 50 mm finder scope, and a 16-inch f6 reflector on a heavily rebuilt equatorial mount with a Meade low profile 2 by 1¼ focuser. The club is active in civic programs, including an Astronomy Days program held at a local mall in May, and is willing to do astronomy programs in local schools and with groups such as the Scouts.

Dues: $15 annual

Delaware Astronomical Society

P.O. Box 652
Wilmington, DE 19899
Phone: (302) 764-1926
E-mail: hbouchel@udel.edu
Members: 150

The Delaware Astronomical Society (DAS), a member-society of the Astronomical League, normally holds meetings at 8 p.m. on the third Tuesday of each month (September through June) at Mount Cuba Astronomical Observatory, located off Hillside Mill Road in Greenville, Delaware. All meetings are open to the public. Special events may be scheduled throughout the year to augment or replace regular meetings. These events include speakers, regional field trips, and dinner meetings with invited speakers.

Members receive *Focus*, the Society's monthly newsletter, as well as the Astronomical League newsletter, the *Reflector.* Other benefits include use of the Society's loaner telescopes, use of its permanently mounted 12.5-inch reflecting telescope, access to the club library, and discounts on many astronomical publications.

Dues: $15 annual

Eastern Michigan University Astronomy Club

Sherzer Observatory
Eastern Michigan University
Ypsilanti, MI 48197
Phone: (313) 487-4146
Fax: (313) 487-0989

The Eastern Michigan University (EMU) Astronomy Club operates the Sherzer Observatory (see separate listing under *Places of Interest*) and holds meetings on Thursday evenings at 7:30 p.m. in 402 Sherzer. The club sponsors open houses at the observatory and observing sessions throughout the academic year (reduced summer schedule). Sessions are free and open to students and the general public on clear Monday evenings, 7:30 p.m. to 10 p.m., and Thursday evenings, 8:30 p.m. to 10 p.m. School and activity groups can arrange for tours by appointment.

El Paso Astronomy Club

Cory Stone, President
El Paso, TX
Newsletter Editor: (915) 833-6948
(Dave Hicks-Klund)
Fax: (915) 833-6527
E-mail: hicksklu@ix.netcom.com
Internet: www.dzn.com/dskies.htm
Members: 65+

El Paso Astronomy Club was founded in August 1995 and meets on the third Friday of every month at the El Paso School District Planetarium (6351 Boeing Drive, El Paso, Texas)

at 7 p.m. Meetings last approximately 1 hour. In addition to regular meetings, the club gathers once a month on the Friday closest to the New Moon for dark sky observing (places vary). Public viewing events are held quarterly and for other special events (i.e. comet watch, lunar eclipses, etc.). The club also sponsors a Messier Marathon in March and a Star-B-Que in October or November. *Desert Skies*, the monthly newsletter of the El Paso Astronomy Club, published its first issue in May, 1996. Back issues can be purchased for $1.00 each.

Dues: $14 annual (individual)
       $17 annual (family)

Escambia Amateur Astronomers' Association

c/o Physical Sciences, Pensacola Junior College
1000 College Blvd.
Pensacola, FL 32504-8998
Phone: (850) 484-1152 (voicemail)
Fax: (850) 484-1822
E-mail: wwooten@pjc.cc.fl.us
Internet: www.meteor.pen.net
Members: 200

since 1976 the EAAA has hosted Sky Interpretation Sessions at Ft. Pickens National Seashore

The Escambia Amateur Astronomers' Association (EAAA) is an active, family oriented astronomy club serving amateur astronomers in northwest Florida and southern Alabama. The Association publishes a monthly newsletter, the *Meteor*, with a circulation of about 500 copies. It holds monthly meetings and programs at the Science and Space Theater of Pensacola Junior College on the Friday closest to the full moon, and monthly deep sky gazes at Munson, Florida on new moon weekends. Every summer since 1976 the EAAA has hosted Sky Interpretation Sessions at Ft. Pickens National Seashore. They hold Astronomy Day activities in conjunction with Earth Day in downtown Seville Square in Pensacola in the spring. The Association also hosts school and Scout stargazes, plus special comet watches and eclipse parties when needed.

European Space Agency

8-10, rue Mario Nikis
F-75738 Paris Cedex 15
France
Phone: (33) 1 5369 7654
Fax: (33) 1 5369 7560

The European Space Agency (ESA) has its roots in the European Launcher Development Organization (ELDO), founded in 1962 by Belgium, France, Germany, Italy, the Netherlands, and the United Kingdom, and the European Space Research Organization (ESRO), founded in 1962 by the same countries, plus Denmark, Spain, Sweden and Switzerland. These organizations merged in 1973 to form the ESA. Austria, Norway, Finland and Canada have joined since the merger, bringing membership to fourteen.

The European Space Agency has launched twelve scientific satellites and probes, making major advance in

Photo: ESA

Artist's impression of the ESA scientific spaceprobe HUYGENS starting its descent through the thick methane atmosphere of the Saturn Moon Titan. The joint ESA/NASA CASSINI/HUYGENS mission will be launched in 1997 and will enter the Saturn planetary system in October 2004.

our knowledge of the universe. These satellites and probes included Giotto (a space probe that examined the nucleus of Halley's Comet, among other things), and the International Ultraviolet Explorer (IUE) observatory, . Of the twelve launched, six are still operational.

European Space Agency programs include Hipparcos (mission: determining the astrometric parameters of stars with unprecedented precision), the Hubble Space Telescope (a joint NASA/ESA program), Ulysses space probe (a joint NASA/ESA program to study the poles of the sun), the Infrared Space Observatory (ISO), the Solar and Helispheric Observatory (SOHO), and the Cassini/Huygens program (a joint NASA/ESA program with the mission of exploring the whole Saturnian system). For a complete list of programs and projects, contact the ESA

Evansville Astronomical Society

P.O. Box 3474
Evansville, IN 47733
Phone: (812) 922-5681
Internet: www.evansville.net/~eas

Evansville Astronomical Society (EAS) has as its primary goal the advancement of amateur astronomy. Founded in 1952, the Society holds regular meetings on the third Friday of each month at 7:30 p.m. at the Wahnsiedler Observatory. This observatory, which is owned and operated by the Society, is located on the grounds of Lynnville Park near the town of Lynnville, Indiana. Lynnville is about 20 miles northeast of Evansville. Built in 1980, this location serves as the organization's headquarters. The observatory contains a lecture hall, a computer room, a photographic dark room, a lounge and a dome housing two telescopes: a 14" Schmitt-Cassegrain reflector and a 9" refractor on a clock-driven, computerized, equatorial mount. There is also a 6" solar scope available.

The Society sponsors public Star Watches and open house events monthly through the warm months and sometimes bi-monthly during the cold months. These events give the public free access to both the Observatory and to the members of the EAS. The club also holds special events during astronomical events such as comets, eclipses, and meteor showers. The Evansville Astronomical Society is a member of the Astronomical League. Members receive the club's monthly publication, *The Observer*, and the quarterly publication of the Astronomical League, the *Reflector*.

Dues: $28 annual (individual)
$34 annual (family)

Friends of Astronomy On-Line

8191 Woodland Shore, Lot #12
Brighton, MI 48114-9310
Phone: (810) 227-9347
E-mail: gary-anderson@usa.net
Internet: www.geocities.com/
Area51/2591

Friends of Astronomy On-Line is a cyberspace club that does everything over the Internet. Members can submit articles and/or photos to the on-line site, sharing their views, ideas, and photos to the Internet community. The club plans on providing an up-to-date web page where children, adults, and educators can access astronomical information.

Galaxies Astronomy Club

Toney Burkhart, Founder
E-mail: galaxies@delphi.com
Internet: www.delphi.com/astro/
index.html
Members: 1485

Galaxies Astronomy Club is the first virtual astronomy club on the Internet. They have over 1485 members and are growing fast. Check them out the next time you're on line.

Greenbelt Astronomy Club

c/o Owens Science Center
9601 Greenbelt Road
Lanham, MD 20706
Phone: (301) 441-4605
Fax: (301) 918-8753
E-mail:
bridgman@grossc.gsfc.nasa.gov
Internet: ssdoo.gsfc.nasa.gov:80/
hbowens/gac.html
Members: 25

Star parties are held at Northway Fields in Greenbelt about once a month

The Greenbelt Astronomy Club meets the last Thursday of each month (except holidays) at 7:30 p.m. at the Howard B. Owens Science Center (call the Science Center at 301-918-8750 to confirm meeting dates). Star parties are held at Northway Fields in Greenbelt about once a month, usually on a Saturday night near the new moon or first quarter moon. Club meetings and star parties are open to the general public. Membership privileges include the newsletter *Meteor*. The Greenbelt Astronomy Club is a member of the Astronomical League and the International Dark Sky Association.

Dues: $12 annual (individual)
$18 annual (family)

Group 70, Inc.

1689 Abram Ct.
San Leandero, CA 94577
Phone: (510) 351-1288
E-mail (Membership):
epoch@majornet.com
Internet: www.hpl.hp.com/astro/
group70/

The group began in 1988 with the goal of building the Large Amateur Telescope, a 72" (1.8M) astronomical instrument.

Group 70 is a non-profit educational organization comprised of people from several countries and many walks of life. The group began in 1988 with the goal of building the Large Amateur Telescope, a 72" (1.8M) astronomical instrument. Upon completion it will be the largest telescope in the world built by and for those who do not have access to large institution-run observatories. The project has grown to not just providing a large aperture telescope, but to offering related instrumentation and services to those in amateur, professional and educational fields of astronomy.

Harford County Astronomical Society

P.O. Box 906
Bel Air, MD 21014
Phone: (410) 836-4155
Internet: www.access.digex.net/
~schapman/hcas.html
Members: 55

The Harford County Astronomical Society holds open houses and observing sessions for the general public at the Harford County Observatory, located on campus at Harford Community College. The club and college jointly operate all astronomy programs and classes.

Photo: Larry Hubble

This photo of Comet Hale-Bopp was taken by Harford County Astronomical Society member, Larry Hubble, using an 11" Celestron telescope, hypersensitive film exposed for 7 minutes, a 35mm camera and a 150mm lens.

Harvard-Smithsonian Center for Astrophysics

60 Garden Street
Cambridge, MA 02138
Phone: (617) 495-7461
Astronomy Info: (617) 496-STAR

Public observing nights are held at the center the third Thursday of each month at 8 p.m. in Phillips Auditorium.

The Center for Astrophysics is a joint facility that combines the resources of the Smithsonian Astrophysical Observatory and the Harvard College Observatory. The Smithsonian Astrophysical Observatory (SAO) was founded in 1890, and the Harvard College Observatory (HCO) was founded in 1839. The prestigious institutions now cooperate in broad programs of astrophysical research that pull on the talents of some 200 Smithsonian and Harvard scientists. Scientific investigations are organized into the following divisions: Atomic and Molecular Physics, High Energy Astrophysics, Optical and Infrared Astronomy, Planetary Sciences, Radio and Geoastronomy, Solar and Stellar Physics, and Theoretical Astrophysics. Major achievements include pioneering the development of instrumentation for orbiting observatories in space, pioneering ground-based gamma-ray astronomy, and more. The Center's observing facilities include the Whipple Observatory in Amado, Arizona, and the Oak Ridge Observatory in Harvard, Massachusetts.

Public observing nights are held at the center the third Thursday of each month at 8 p.m. in Phillips Auditorium. The Center also serves as the headquarters for the Central Bureau for Astronomical Telegrams and the Minor Planet Center of the International Astronomical Union.

Hopatcong Area Amateur Astronomers

P.O. Box 360
Ledgewood, NJ 07852-0360
Phone: (201) 579-6453 (Bob)
**   or (201) 770-1625 (Ken/Karen)**
E-mail: jeffer@cdsusa.com
Members: 30

The Hopatcong Area Amateur Astronomers (HAAA) is a club for people of all ages who share a common interest in astronomy. Founded in 1990 by local residents Duane Degutis and Ken Roundy, most of the club's members are from Morris and Sussex counties. The club meets the third Wednesday of every month at 7:30 p.m. at the Hopatcong Civic Center on Lakeside Boulevard. Some members are involved in astrophotography, solar observing, and CCD imaging. Visitors are welcome to attend monthly business meetings and observing sessions afterwards, when the skies are clear. The club's goal is to share their enthusiasm and expertise in astronomy among members and the public.

HAAA is a member of the United Astronomy Clubs of NJ (UACNJ), and has access to the observatory at Jenny Jump State Forest, located near Hope, New Jersey. They participate in state-wide astronomy events and in astronomical conventions in other states, and occasionally hold educational workshops for local public and private schools, Scouting groups and summer camps.

Dues: $15 annual (first year)
        $20 annual (after first year)

International Association for Astronomical Art

Dale Darby, Membership Director
6248 Carl Sandburg Circle
Sacramento, CA 95842
Phone: (916) 331-0147
E-mail: darby0147@aol.com
Internet: www.novaspace.com/
IAAA/IAAA.shtml
Members: 140

The International Association for Astronomical Art (IAAA) includes everyone from astronauts (Alan Bean of Apollo and Gemini missions) to collectors and, of course, the finest space artists on the planet. Many of the organization's artists have had their work presented in movies and books. The IAAA's goal is to educate the public on astronomy and to promote space ventures. They do a lot of work for the Planetary Society, NASA and PBS.

International Dark-Sky Association

3545 N. Stewart Avenue
Tucson, AZ 85716-1241
Phone: (520) 293-3198
E-mail: SaveOurSky@aol.com
Internet: www.darksky.org
Members: 2100

The goal of the International Dark-Sky Association (IDA) is to stop the adverse environmental impact on dark skies created by light pollution, and to address the problems of radio interference and space debris that affect the work of most existing observatories. One of the ways the association accomplishes its goal is by educating the public regarding quality lighting. In fact, a number of lighting companies are among the organizations 2100 members from 50 states and 70 countries.

Dues: $30 annual (individual)
        $15 annual (student)

International Meteor Organization

161 Vance Street
Chula Vista, CA 91910-4828
E-mail:
LUNRO.IMO.USA@prodigy.com
Internet: www.imo.net
Members: 250 worldwide

Founded in 1988, the International Meteor Organization (IMO) is an international scientific non-profit organization with worldwide membership. The organization was created in response to an ever growing need for international cooperation of amateur meteor work. As such, the IMO's main objectives are to encourage, support, and coordinate meteor observing; to improve the quality of amateur observations; to disseminate observations and results to other amateurs and professionals; and to make global analyzes of observations received worldwide.

The primary instrument used by IMO to achieve these goals is its bimonthly journal *WGN*, which all members receive. Annually, this journal contains over 220 pages of general meteor news, observing program guidelines, reports, analyzes of observations, and more general articles on meteoric phenomena.

There are several active commissions within the IMO. These are divided into the following categories: visual, fireball, photographic, and telescopic. Commissions for radio and video will soon be formed. Direct links to each commission can be provided by the regional coordinator. The IMO has published three books which may be ordered from the regional coordinator, they are *Photographic Astrometry* by Christian Steyaert, *Handbook for Visual Meteor Observers* by Jurgen Rendtel and Rainer Arlt, and *Handbook for Photographic Meteor Observations* by Jurgen Rendtel.

Dues: $25 annual (surface)
        $50 annual (airmail)

The Jackson (Hole) Astronomy Club

Walt Farmer, Chair
P.O. Box 1821
Jackson, WY 83001
Phone: (307) 733-2173
Fax: (307) 733-2173
E-mail: astrocowboy@juno.com
Members: 37

Joint Astronomy Centre

660 N. A`ohoku Place
University Park
Hilo, HI 96720
Phone: (808) 961-3756
Fax: (808) 961-6516

The Joint Astronomy Centre is responsible for operating the James Clerk Maxwell Telescope (JCMT) and the United Kingdom Infrared Telescope (UKIRT). These telescopes are located on top of Mauna Kea, a 4200 meter dormant volcano on the Big Island of Hawaii.

The James Clerk Maxwell Telescope is funded by the United Kingdom, Canada, and the Netherlands, although astronomers from any country can apply to use the telescope. The United Kingdom Infrared Telescope is also funded by the United Kingdom, but used internationally.

The JCMT and UKIRT are not open to the public, but a daytime visit to the summit allows the many telescopes to be seen from the outside, and gives spectacular views of the island. Note: A four-wheel-drive vehicle is essential. The easiest way to visit is to arrange a tour with a local sightseeing operator.

Kalamazoo Astronomical Society

c/o KAMSC
600 West Vine, Suite 400
Kalamazoo, MI 49008
Phone: (616) 337-0004
E-mail: stargazer@voyager.net
Internet: www.geocities.com/
CapeCanaveral/Lab/1000/kas.html
Members: approximately 70

The Kalamazoo Astronomical Society was found in 1936, making it the second oldest club in the state of Michigan. General Meetings are held on the first Wednesday of every month at 7 p.m. (except holiday weekends when meeting is delayed one week) in the Universe Theater & Planetarium located at the Kalamazoo Valley Museum (230 North Rose Street, Kalamazoo, MI 49007). Meeting sites can vary, so please visit the KAS Home Page or call before coming.

Membership benefits include monthly observing sessions, loan of club telescopes and equipment, subscription to *Prime Focus* newsletter, lectures, field trips, magazine discounts, access to club library, membership in the Astronomical League, use of club observatory (when completed), and more.

Dues: $20 annual (Individual)
        $25 annual (Family)
        $15 annual (Students/Seniors)

Kanawha Valley Astronomical Society, Inc.

Rodney F. Waugh
5292 Questa Drive
Cross Lanes, WV 25313
Phone: (304) 776-2613
E-mail: rwaughkvas@aol.com

The Kanawha Valley Astronomical Society (KVAS) usually meets the third Friday of each month at 7:30 p.m. at Mathes Hall behind St. Matthews Church. The Society maintains the Breezy Point Observatory at Camp Virgil Tate (the observatory houses a 12 1/2" Newtonian telescope). KVAS works closely with Camp Virgil Tate, Sunrise Museum, schools, and other groups and institutions to provide astronomical programs and star parties for people of all ages.

Membership benefits include a monthly newsletter, use of Breezy Point Observatory and the club library, and magazine discounts.

Dues: $20 annual (individual/family)

Kern Astronomical Society

712 Sesnon Street
Bakersfield, CA 93309
Phone: (805) 393-4387 (President)
Phone: (805) 397-0327 (Treasurer)
E-Mail (President):
102163.540@CompuServe.Com
Internet: users.aol.com/frankrip/
kas/kas.html
Members: 100

The Kern Astronomical Society covers central Kern County, which is located at the southern end of the San Joaquin Valley in California. The Society meets in Bakersfield where most of the members live. The Kern Astronomical Society promotes community awareness of current astronomical events and provides equipment for public groups to observe the heavens. A special effort is made to introduce astronomy to the community's children, including opportunities to observe. The Society meets on the first Friday of each month, except July, at 7:30 p.m. in the Northminster Presbyterian Church Social Hall, 3700 N. Union Ave. in Bakersfield. The July meeting is a picnic outing. One or more star parties are held monthly, usually in the mountains outside Bakersfield, as weather permits. Members receive a monthly newsletter.

Dues: $12 annual

Kokomo Astronomy Club

2300 S. Washington Street
P.O. Box 9003
Kokomo, IN 46904-9003
Phone: (765) 455-9305
Fax: (765) 455-9528
E-mail:
fsteldt@iukfs1.iuk.indiana.edu
Members: 15

The Kokomo Astronomy Club meets at 7 p.m. on the second Sunday of each month from September through May at the Indiana University Kokomo Observatory. Club members have the opportunity to take part events such as a trip to Maracaibo, Venezuela to view the February 26, 1998 total solar eclipse. The Indiana University Kokomo (IUK) Observatory houses a permanently mounted Celestron-14 telescope in a 15-foot Ash Dome. The dome includes a Lanphier Shutter System with an observing window. This is the only public observatory in the state of Indiana with this feature, so the dome can be heated in the winter.

Dues: $10 annual
        $5 annual (students)

Lackawanna Astronomical Society

1112 Fairview Road
Clarks Summit, PA 18411
Phone: (717) 586-0789
E-mail: sabiajohn@aol.com
Internet: members.aol.com/
sabiajohn/LAS.html
Members: 110

The Lackawanna Astronomical Society holds regular meetings on the first Tuesday of each month. The meetings are usually held at Keystone Observatory located in Fleetville, Pennsylvania. During meetings, members of the Society can make use of the telescopes owned by the Society and the instruments owned by Keystone College. The Society has constructed a roll-off roof Observatory for their 12.5 inch f/5.6 home made reflecting telescope. This is located on the grounds of Keystone Observatory and can be used by Society members. The scope is on a clock driven equatorial mount.

Club Night at Keystone Observatory is held the Saturday night after the regular meeting. This is an activity where members may use the telescopes at the observatory or their own instruments at the observatory's dark sky location. Observing sessions during special astronomical events are also sponsored by the Society. These can include public viewing of meteor showers, comets, lunar eclipses, or other phenomena. The date and time of these events are published in *The Ecliptic*, the Society's newsletter. During the summer months, the Society sponsors a slide program and observing sessions (using member-owned telescopes) at Promised Land State Park and Lackawanna State Park. Another event the Society organizes each year is Astronomy Day, a public

> *The Society has constructed a roll-off roof Observatory for their 12.5 inch f/5.6 home made reflecting telescope.*

event where displays, programs and information on astronomy and current astronomical events are presented. Viewing of the night sky with binoculars and telescopes is also done if weather conditions permit. There are no fees or charges for any of the events described. Many club members also act as volunteers and assistants at the Public Nights, Open House evenings, and Lecture Series given by Keystone College, which owns and operates the observatory.

Membership benefits include a subscription to *The Ecliptic*, discounts on astronomical publications, discounts on other items and publications, use of the club's 6-inch reflector telescope, and access to the Society's library.

Lake County Astronomical Society

603 Dawes
Libertyville, IL 60048
E-mail: jak.stargate@worldnet.att.com
Internet: homepage.interaccess.com/
~purcellm/
Members: 60

The Lake County Astronomical Society (LCAS) has members who are accomplished observers and astrophotographers, and others who are new to astronomy. Some own large telescopes, while others simply enjoy

observing the night sky with their unaided eyes. The club tries to respond to all levels of interest. Activities include club meetings the third Friday of each month (at Libertyville main fire station on the west side of Milwaukee Avenue, just south of Route 137), field trips to planetaria and observatories, star parties, observing sessions at LCAS dark-sky sites, and annual trips to the southwest. The Society publishes a monthy newsletter, *Night Times*.

Dues: $8 annual (individual)
              $12 annual (family)

League of the New Worlds

P.O. Box 542327
Merritt Island, FL 32954
Phone: (407) 634-5151
E-mail: league@aol.com
Internet: members.aol.com/league/
index.html

The League of the New Worlds is the only group in the world organized to actively seek out and train tomorrow's colonists of space and the ocean floor. They are building the actual components of space and ocean colonies, using the ocean as an analog to space. They are a very serious group of trained and dedicated exploration professionals backed up by a team of astronauts (Scott Carpenter), aquanauts (Dennis Chamberland) and space engineers (NASA engineer Richard Schealer) as well as teachers, students and other interested citizens around the world.

Annually, the organization mails out five editions of *The New Worlds Colonist*, a copy of their annual professional journal, *The New Worlds Explorer*, and frequent e-mail updates to their members.

Dues: $35 annual

Lehigh Valley Amateur Astronomical Society

East Rock Road
Allentown, PA 18103
Phone: (610) 797-3476
Internet: www.lvaas.org
Members: 240

The Lehigh Valley Amateur Astronomical Society, Inc. (LVAAS) is a public-oriented nonprofit educational organization dedicated to serving the interests of the community in astronomy and related fields. Established in 1957 by local amateurs in the Allentown-Bethlehem area, the Society has been in continuous operation for four decades. Totally self-supporting, LVAAS flourishes through its memberships voluntary contributions of talent and skill.

The Society is a meeting place for people with a wide range of interests and backgrounds. Some are serious observers; some just observe for fun. Others study to become active contributors to astronomy-related periodicals, or to present lectures or planetarium programs. Whatever the occupation, educational background, or interest, the Society has much to offer through its many programs and facilities.

The Society's administrative headquarters and three observatories are located on one acre atop South Mountain. The main building houses a meeting room, the Grady Planetarium, the Robson Library/Computer Wing, a workshop, and an optical room. The planetarium seats about forty people, under a 21-foot dome, that has the unique feature of descending to eye level, to simulate a realistic horizon. The library has accumulated an extensive collection of astronomically related material. This has been recently expanded to include videos, which are available for borrowing by Society members.

The observatories atop South Mountain feature a 6" F/15 refractor, a 12.5" F/16 Cassegrain and a 12.5" F/6 Newtonian reflector. Located in the main building are numerous portable instruments which are available on a monthly basis for a small rental fee to members. The Society's dark sky site is located 25 miles west of Allentown between Lenhartsville and Hamburg, Pennsylvania. Known as Pulpit Rock Astronomical Park, or as it is commonly called, "The Rock," this excellent mountain top site sits 1,600 feet above sea level near the Appalachian Trail. Telescopes at Pulpit Rock include an 8" F/15 Space refractor, a 20" F/16 Schmidt-Cassegrain reflector, a 12.5" F/7 Newtonian reflector and a 17.5" Dobsonian. The Society's newest project at the "Rock" is a microcomputer controlled, 40" F/16 Cassegrain telescope to be completed by 1999.

The installations and equipment at Pulpit Rock offer the serious amateur or the novice an opportunity to contribute meaningful scientific information to the astronomical community or to simply view the splendors of the heavens from two acres of landscaped grounds.

For a membership application, please contact: LVAAS Membership Records, c/o Robert Mohr, P.O. Box 368, Fogelsville, PA 18051-0368, phone (610) 398-7295.

Dues: $35 annual

The LodeStar Project

Institute for Astrophysics
University of New Mexico
Department of Physics & Astronomy
800 Yale Blvd., NE
Albuquerque, NM 87131-1156
Phone: (505) 277-4307
Fax: (505) 277-9657
E-mail:
lodestar@lodestar.phys.unm.edu
Internet: lodestar.phys.unm.edu

The ambitious LodeStar Project is an astronomy educational initiative that combines education, teacher training, public outreach, research and tourism. Led by the University of New Mexico (UNM), the LodeStar Project will ultimately consist of three facilities; a computer-controlled telescope site at the New Mexico Institute of Mining and Technology, a hands-on learning center in Albuquerque, and the Enchanted Skies Park in Cibola County - the first park in the nation dedicated to public stargazing.

Made possible by an Air Force Office of Scientific Research grant for $15.8 million and $12 million in state funds, the LodeStar Project is designed to improve math and science education in New Mexico, provide teacher training, accommodate the research efforts of professional and amateur astronomers, and advance the technological base of the nation.

The Socorro site observatory at the New Mexico Institute of Mining and Technology, the first of the three sites to open, provides a computer link-up between schools statewide and a telescope at the institute. It will also feature a 30-inch research telescope on South Baldy Mountain in the Cibola National Forest in the near future.

The hands-on learning center, the second site, will be integrated with the New Mexico Museum of Natural History and Science and will be designed to make astronomy much more accessible to New Mexico's schoolchildren and to the public. The center will be the principal location for teacher training, educational programs, and public outreach and will serve as a local model that may be used in other science programs nationwide.

Finally, LodeStar's Enchanted Skies Park, located in the Grants/ Acoma area of New Mexico (70 miles west of Albuquerque), will be the first park dedicated to the public exploration of the night sky.

The park will feature many telescopes for observers of all ages and a forum for Native Americans to share and sustain their traditional views of the night sky. Also, the park will support a research quality observatory and telescopes which will lead the way into the next century's exploration of the universe. Enchanted Skies Park is scheduled to open in 1999.

Louisville Astronomical Society, Inc.

Charles Allen, Secretary
P.O. Box 701043
Louisville, KY 40270-1043
Phone: (502) 228-3043
Fax: (502) 581-1087
E-mail:
74023.2331@compuserve.com
Internet: www.venus.net/
~dhaggard/las.htm
Members: 75

The Louisville Astronomical Society (LAS) was founded in 1933. Over $40,000 has been raised toward construction of a new LAS observatory. The LAS conducts over 40 public lectures and observations each year including a major annual summer star party at Newton-Stewart State Recreation Area, Indiana (Patoka Lake). Other events include joint picnics with the Evansville Astronomical Society and an annual winter party.

Club members receive the LAS' monthly newsletter, *Starword*, and enjoy the right to rent club telescopes. Meetings are held at 8 p.m. on the first Friday of each month (second Friday to avoid a holiday weekend) in Room 102 Strickler Hall, University of Louisville, Belknap Campus. One or two club observations are held each month at Wyandotte State Recreation Area, Hickory Hollow Nature Center, near Corydon, Indiana.

The LAS will co-host the Astronomical League's national convention in July 1998. Current LAS members include the League's Vice-President, the League's Great Lakes Regional Representative, the Chair and Vice-Chair of the League's National Young Astronomer Award, a winner of the Astronomical League Award (Virginia Lipphard - 1991) and two winners of the Astronomical League Outstanding Junior Award (Daniel Klineman in 1962; J. Richard Gott in 1963). The National Coordinator of the National Deep Sky Observers Society is an LAS member and, in the 1990s alone, seven LAS members have won a total of 35 regional and national awards for astrophotography, CCD imaging and telescope making.

Dues: $25 annual (individual)
$30 annual (family)
$15 annual (senior 65+)

Madison Astronomical Society

Madison, Wisconsin
c/o Bob Manske
404 Prospect Rd.
Waunakee WI, 53597
Phone: (608) 849-5287
Members: 80

The Madison Astronomical Society operates the Yanna Research Station (YRS) about 40 miles south of Madison. The site includes four permanently mounted telescopes: a 17.5-inch Dobsonian, a 16-inch classical Cassegrain, a 12-inch computer driven Schmidt-Cassegrain, and an 11-inch Schmidt-Cassegrain. Three portable telescopes complete the array. There are also concrete pads supplied with electrical power for members who wish to bring their own equipment.

operates the Yanna Research Station

The Society meets at 7:30 p.m. in the Edgewood High School Library on the second Friday of January, February, March, May, July, August, October and November, and on the third Friday in September. The annual banquet is held in April, usually on the second Friday, and the annual picnic is held the second Saturday of June. The Society's annual Christmas party is on the second Friday of December at U.W. Space Place.

Dues: $20 annual (regular)
        $45 annual (observing membership, gives full access to the telescopes at YRS)

Magic Valley Astronomical Society

Forrest Ray, President
531 Filer Avenue, Apt. 3A
Twin Falls, ID 83301
Phone: (208) 736-8678
Members: 19

The Magic Valley Astronomical Society meets the second Saturday of each month at the Jerome Public Library, Jerome, Idaho. Observing sessions follow regular meetings. Observing sessions are also held on the Saturday nearest the new moon. The club is active in mirror making, astrophotography, and they sponsor public star parties. The Society is a member of the Astronomical League.

Dues: $5 annual

Manteca Observing Group

Contact: Pat Lambertson
14533 Chickasaw Way
Manteca, CA 95336
Phone: (510) 782-4322, ext. 334
E-mail:
72170.2453@compuserve.com
Members: 18

The Manteca Observing Group is active in promoting public awareness and education of astronomy. Membership includes several Project Astro Teacher/Astronomer partnerships, a program started and run by the Astronomical Society of the Pacific.

Mauna Kea Astronomical Society

P.O. Box 5628
Kailua-Kona, HI 96745
Phone: (808) 928-8974
E-mail: psears@aloha.net
Members: 40

Miami Valley Astronomical Society

Dayton Museum of Discovery
2600 DeWeese Parkway
Dayton, OH 45414
Phone: (937) 275-7431
Internet: www.mvas.org
Members: 130

The Miami Valley Astronomical Society (MVAS) can trace its roots back to around the time of World War I. Currently headquartered at the Dayton Museum of Discovery, MVAS holds its general meetings on the second Friday of each month (except June) at 7:30 p.m. The Society operates and main- tains three observatories in southwest- ern Ohio, including the Apollo Obser- vatory (AO) located at the Dayton Mu- seum of Discovery, the Junior Observ- ing and Training Station, commonly re- ferred to as the Junior Observatory (also at the Dayton Museum of Dis- covery), and the John Bryan Obser- vatory, located at John Bryan State Park near Yellow Springs, Ohio. The Society hosts the annual Apollo Ren- dezvous, a convention held in June for amateur astronomers from the Great Lakes Region. Member benefits in- clude a club library and *Amateur As- tronomer*, the Society's monthly news- letter.

Midlands Astronomy Club

P.O. Box 2527
Columbia, SC 29202
E-mail: wilson@scuch8.psc.sc.edu
Internet: astr.psc.sc.edu/ htmlpages/MAC/MAC.html
Members: 50

The Midlands Astronomy Club is a dedicated and enthusiastic group of amateur astronomers interested in ob- serving, telescope making, photogra- phy and public education. Headquar- tered in Columbia, South Carolina, they have two main observing sites where they hold monthly star parties. All members and their guests are invited to come out and sample dark skies with a large variety of telescopes. The club has two 20-inch Dobsonians, mul- tiple large aperture refractors, many Schmidt-Cassegrains and medium- sized Newtonians and a smattering of giant binoculars. Long-time member and comet hunter Howard Brewington found his first comet while active in this club.

A subset of the club's members are highly skilled telescope makers. They have ground and polished mirrors up to 12 inches, and have constructed numerous Dobsonian and equatorial Newtonians which are owned by club members. They have also made various mounts for giant binoculars and double-arm barndoor trackers. They have a broad base of skills including expert woodworking, machining and electronics.

Another subset of members are expert astrophotographers, with works published in *Astronomy* and *Sky & Telescope* magazines. Quarterly astrophoto contests are held, and assistance is provided to those members wishing to begin photography or improve their skills.

Besides the monthly star party, the club holds public observing events during astronomical events (they had 2500 people attend their Comet Hale-Bopp event). A pool of volunteers is also available to visit schools to hold education sessions and star parties for parents and students.

Millstream Astronomy Club

Roger L. Myers, President
305 N. Westwood
Lima, OH 45805
E-mail: rogergop@bright.net

Mobile Astronomical Society

c/o Rod Mollise
1207 Selma St.
Mobile, AL 36604
Phone: (334) 432-7071
E-mail: rmollise@aol.com
Internet: members.aol.com/
RMOLLISE/index.html
Members: 40

The Mobile Astronomical Society is fortunate to have the Mobile Public Schools' Environmental Studies Center available for meetings. This beautiful wooded tract of land is within the Mobile city limits and the Center itself features state-of-the-art classroom facilities, and is fully equipped with audio-visual/computing equipment. The Environmental Studies Center (ESC) also allows the Mobile Astronomical Society the use of a plot of land on the ESC grounds for the club's roll-off roof observatory. This facility features a heavy duty Cave equatorial mount capable of handling either of the club's telescopes: a 12.5" f6 Cave Newtonian or an 8" Celestron SCT. In return for the use of ESC facilities, the Mobile Astronomical Society participates in a number of public outreach astronomy programs and public star parties throughout the school year.

Mobile Astronomical Society participates in a number of public outreach astronomy programs and public star parties throughout the school year

Monterey Institute for Research in Astronomy

200 Eighth Street
Marina, CA 93933
Phone: (408) 883-1000
Fax: (408) 883-1031
E-mail: mira@mira.org
Internet: www.mira.org

Founded in 1972, the Monterey Institute for Research in Astronomy (MIRA) is a private, non-profit institute dedicated to astronomical research and education. MIRA astronomers are involved in many research projects, such as their development of a new system that uses artificial intelligence techniques on near-infrared stars to classify them more accurately than by visual observation alone. Education programs include public lectures, star parties, and a new Internet program, *Field Trip to the Stars*. MIRA's observatory, the Oliver Observing Station, sits atop Chews Ridge and features a 36-inch telescope. MIRA has also received four buildings at Fort Ord, a former Army base, which, once remodeled, will be MIRA's new Astronomy Center. Membership in the institute's support organization, the Friends of MIRA, is available to those wishing to support the important work and the expansion of MIRA.

Dues: $35 annual (individual)
        $50 annual (family)
        $15 annual (student)

The Morris Museum Astronomical Society

Morris Museum
6 Normandy Heights Road
Morristown, NJ 07960
Phone: (201) 386-1848
Fax: (201) 386-0969
E-mail: apisano@ix.netcom.com
Members: 75

Founded in 1970, the Morris Museum Astronomical Society (MMAS) meets the second Thursday every month (except July & August) at the Morris Museum from 8 p.m. until 9:30 p.m. Both astronomical and solar observation are done at different times during the month from a facility at the museum.

Dues: $15 annual

Mountain Skies Astronomical Society

P.O. Box 1169
Lake Arrowhead, CA 92352
Phone: (909) 336-1699
24-hour Sky Report: (909) 336-1299
E-mail:stargazersmail@worldnet.att.net
Internet: www.astro-
msas.holowww.com
Members: 2122

Not many astronomical societies can consider opening their own $6.2 million Observatory and Science Center, but that is exactly what Mountain Skies Astronomical Society (MSAS) is doing. The Society already has a Science Center, located in Lake Arrowhead Village Executive Suites. However, the new 12,000 sq. ft. Observatory and Science Center would expand

Art: Mountain Skies Astronomical Society

their ability to meet public education goals by providing a planetarium, observatory, library, conference room, photo lab, and assembly hall. Once complete, this facility will be an outstanding educational asset for the local community and surrounding area.

Mountain Skies Astronomical Society has a Star Atlas program that allows individuals to name a star, either as a gift, for a special occasion, or just for the fun of it (see MSAS listing under *Products & Services* for details).

Membership benefits include a subscription to *Stellar Matters*, the Society's quarterly newsletter, a 10% discount on all MSAS merchandise, monthly sky maps, use of the MSAS Science Center, participation in Star Watches and astronomical workshops, a membership card, and more.

Dues: $30 annual (individual
$40 annual (family)
$20 annual (student)

Mutual UFO Network, Inc.

103 Oldtowne Road
Seguin, TX 78155-4099
Phone: (830) 379-9216
Fax: (830) 372-9439
UFO Hotline: 1-800-UFO-2166
E-mail: mufonhq@aol.com
Internet: mufon.com
MUFONET-BBS: (901) 327-1008
MUFON Amateur Radio Net: 40 meter
- 7.237 MHZ, Saturday 8 a.m. EST
Members: 5000+

The Mutual UFO Network (MUFON) is an international scientific organization composed of people seriously interested in studying and researching unidentified flying objects (UFOs). Founded on May 31, 1969, MUFON is governed by a Board of Directors composed of 21 men and women and administered by an Executive Committee. MUFON's Board of Consultants, most of whom are PhDs or MDs representing 45 areas of science, technology, medicine, psychiatry, psychology, theology, engineering, astronomy, communications, political science, photo analysis, etc., are readily available as an advisory group to apply their expertise to UFO cases. The organization maintains a UFO hotline where UFO sightings can be reported.

Membership in the Mutual UFO Network is by invitation only, however, anyone interested in helping to resolve the UFO phenomenon is invited to submit a membership application and appropriate dues for approval. Based upon their education and experience, members may serve in one or more of the following positions: Consultant, State or Provincial Director, State Sec-

tion Director, National Director, Foreign Representative, Field Investigator, Research Specialist, Amateur Radio Operator, Astronomy, Field Investigator Trainee, Translator, UFO News Clipping Service, Journal Subscriber or Associate Member (under 18 years of age).

Membership in the Mutual UFO Network includes a subscription to the twenty-four page monthly magazine, *MUFON UFO Journal*, which is the organizations most significant means of sharing details of UFO sighting reports and vital information related to the UFO phenomenon.

Dues: $30 annual

NACOMM

Tim Hagemeister, Director
P.O. Box 511
Anoka, MN 55303-0511
Phone: (612) 753-5146
E-mail: comments@nacomm.org
Internet: www.nacomm.org

NACOMM (which stands for New Age Communications) is an organiza-

tion devoted to the research and investigation of Unidentified Flying Objects and other related phenomenon. Its emphasis is on the moral implications of intelligent extraterrestrial life, as well as truth in government. NACOMM features up-to-date UFO and aerospace news as well as articles based on Biblical and theological perspectives.

NASA Headquarters

300 E Street, S.W.
Washington, D.C. 20546
Phone: (202) 358-0000
Internet: www.nasa.gov

The National Aeronautics and Space Administration (NASA) Headquarters has more than 2,000 employees and is responsible for the management of the space flight centers, research centers and other installations that constitute the National Aeronautics and Space Administration.

NASA activities are directed by a series of specialized offices, which are described below.

The Office of Aeronautics directs the agency's aeronautics research and development programs, including the High-Speed Research Program which is creating and refining the technology and addressing the environmental challenges supporting the development of a future U.S. high-speed civil transport aircraft.

The office also researches advanced technology for subsonic aircraft, manages NASA's weather-related flight safety research, works to improve inspection methods for aging aircraft, propulsion research and development of advanced piloting and air traffic control aids. In addition, it directs numerous flight research programs using high-performance aircraft such as the SR-71, F/A-18 and F-16XL. It also manages fundamental aeronautics research in aerodynamics, fluid dynamics, structural mechanics and human factors issues such as the interaction of pilots with highly-automated cockpits.

The aeronautics office also manages NASA's portion of the multiagency High Performance Computing and Communications program, and NASA's part of the National Aero-Space Plane (NASP) program. NASP is a national endeavor to develop and demonstrate technology for advanced vehicles that would take off horizontally, fly into orbit, then return for a runway landing.

The Office of Aeronautics has institutional management responsibility for Ames Research Center, Mountain View, California; Ames-Dryden Flight Research Facility, Edwards, California; Langley Research Center, Hampton, Virginia; and Lewis Research Center, Cleveland, Ohio. Dr. Wesley L. Harris is Associate Administrator.

The Office of Space Science is responsible for the NASA space research and flight programs directed toward scientific investigations of the solar system and astronomical objects using ground-based, airborne and space technologies including sounding rockets and deep space satellites. This office works closely with the scientific community through the Space Studies board of the National Academy of Sciences and other advisory groups.

The Office of Space Science has institutional management responsibility for the Jet Propulsion Laboratory, Pasadena, California. Dr. Wesley T. Huntress, Jr., is the Associate Administrator.

The Office of Mission to Planet Earth is responsible for NASA's Earth science and environmental research. Mission To Planet Earth is a comprehensive, coordinated research program that studies the Earth as a global environmental system. Comprising ground-based, airborne and space-based programs, this office includes participation from other federal agencies as part of the U.S. Global Change Research Program and the international science community. The office has institutional management for the Goddard Space Flight Center, Greenbelt, Maryland. Dr. Shelby G. Tilford is Acting Associate Administrator.

The Office of Life and Microgravity Sciences and Applications is responsible for assuring the health and safety of humans in space and to understand the biological effects of space flight on organisms. It also uses the unique attributes of the space environment to conduct research and gain new knowledge in fluid behavior, combustion science, material science and biotechnology. Dr. Harry Holloway is the Associate Administrator.

The Office of Space Flight operates the Space Shuttle and develops both manned and unmanned platforms which enable scientific research and advanced technology development. The Space Shuttle is NASA's primary space transportation system and the only space vehicle capable of carrying people and large payloads into Earth orbit and returning them. OSF is responsible for scheduling Space

Shuttle flights, developing financial plans and pricing structures and providing services to users. As part of its duties, the Office of Space Flight conducts operations and utilization of Spacelab, a laboratory dedicated to research in space that flies in the Shuttle's cargo bay.

The office is working with the Russian Space Agency to plan and execute a series of joint missions that will involve flying a cosmonaut aboard the Shuttle and an astronaut aboard the Mir space station, leading up to a mission with a Shuttle docking to the Russian space station. The office also is conducting early planning activities for the operation of the U.S. space station.

The Office of Space Flight also is responsible for institutional management of the Kennedy Space Center, Florida; Marshall Space Flight Center, Huntsville, Alabama; Johnson Space Center, Houston, Texas; and the Stennis Space Center near Bay St. Louis, Mississippi. Jeremiah W. Pearson III is Associate Administrator.

The Office of Space Systems Development is responsible for defining and developing potential future space systems and capabilities, as well as demonstrating enhancements to improve existing systems capabilities. The office has responsibility for space station development and operations; large propulsion systems development including a new space transportation main engine and the Advanced Solid Rocket Motor (ASRM) and advanced transportation systems program planning.

A permanently manned space station is essential for advancing human exploration of space. The space station will be a permanent outpost in space where humans will live and work productively for extended periods of time. It will provide an advanced research laboratory to explore space and employ its resources, and will provide the opportunity to learn to build, operate and maintain systems in space. The station will be launched in segments aboard the Space Shuttle and assembled in orbit, with first flight set for 1996. NASA centers responsible for developing major elements of the space station are the Marshall Space Flight Center, Johnson Space Center and Lewis Research Center.

The Advanced Solid Rocket Motor is being developed to replace the redesigned solid rocket motor. The ASRM will improve the safety, reliability and the performance of the Space Shuttle system. Arnold D. Aldrich is Associate Administrator.

The Office of Advanced Concepts and Technology has a mission to pioneer innovative, customer-focused concepts and technologies, leveraged through industrial, academic and government alliances, to ensure U.S. commercial competitiveness and preeminence in space.

The office's four primary functions are to maintain a highly professional systems engineering team capable of detailed feasibility and cost analysis of advanced concepts, to be NASA's front door to businesses which want the agency's help and expertise in developing new ideas and technologies, to be the agency's lead in the transfer of technology into the commercial sector and to further the commercialization of space.

The office also manages the agency's Small Business Innovative Research, technology transfer, Defense Conversion Act and other innovative technology development programs including a new experiment in incubating technology start-up companies. Gregory M. Reck is Associate Administrator.

The Office of Space Communications is responsible for planning, development and operation of worldwide communications, command, navigation and control, data acquisition, telemetry and data processing essential to the success of NASA programs and activities.

Communications systems requirements for Space Shuttle flights; Earth orbital, planetary and interplanetary space probes; expendable launch vehicles; research aircraft; sounding rockets; balloons and administrative support are provided by this office. The office consists of five divisions. Charles T. Force is Associate Administrator.

The Office of Safety And Mission Quality plans, develops and evaluates safety, quality and risk management policies and activities in support of NASA programs. Responsibilities include providing leadership in quality management for science and engineering programs and working closely with NASA flight, ground operations and research programs to develop safety, reliability, maintainability and quality assurance policies and requirements. The office consists of seven divisions and three safety panels. Frederick D. Gregory is Associate Administrator.

NOTE: The above information was extracted directly from NASA's exceptional web site and is gratefully acknowledged. For information on NASA sites, such as Kennedy Space Center, Marshall Space Flight Center or Johnson Space Center, see separate listings in the *Places of Interest* section. For information on the Johnson Space Center Visitor Center, refer to the Space Center Houston listing in *Places of Interest*.

National Capital Astronomers

10500 Rockville Pike, Apt. M10
Rockville, MD 20852
Phone: (301) 564-6061
or (202) 966-0739
Members: 250

Established in 1937, the National Capital Astronomers (NCA) is dedicated to advancing space technology, astronomy, and related sciences through information, participation, and inspiration. The organization pursues this goal via research, lectures and presentations, publications, expeditions, tours, public interpretation, and education. NCA is the astronomy affiliate of the Washington Academy of Sciences and holds regular monthly meetings September through June. All are welcome to join. NCA publishes a monthly newsletter called *Star Dust* which is available to members. In addition to announcement of upcoming NCA meetings and a calendar of monthly events, *Star Dust* contains reviews of past meetings and articles on current astronomical events.

The NCA holds mirror-making classes every Tuesday evening from 7 p.m. to 9:30 p.m. at the Chevy Chase Community Center at Connecticut Avenue and McKinley Street, N.W. in Washington D.C. For more information, call (202) 362-8872.

The National Deep Sky Observers Society

Alan Goldstein, National Coordinator
1607 Washington Blvd.
Louisville, KY 40242-3539
Phone: (502) 426-4399
E-mail: Deepskyspy@aol.com
Internet: www.erols.com/njastro/
orgs/ndsos.htm

The National Deep Sky Observers Society (NDSOS) was founded in 1976 to promote deep sky observing and serve as an information resource for deep sky observers. The NDSOS connects distant observers through its membership directory of regular and e-mail addresses. With an arsenal of specialized catalogs, atlases and other resources, the "Data Coordinator" can provide technical information, identify "mystery" objects, and is ready to assist deep sky observers with their personal observing programs. Members receive *Betelgeuse*, the journal of the NDSOS, and *The Practical Observer*, a publication that focuses on techniques, equipment, opinions, product reviews, Internet sites, and other aspects of astronomy as a hobby.

Dues: $20 annual
$23 annual (Canada/Mexico)
$25 annual (other foreign)

The National Space Society

600 Pennsylvania Avenue, S.E.
Suite 201
Washington, D.C. 20003-4316
Phone: (202) 543-1900
Fax: (202) 546-4189
E-mail: nsshq@nss.org
Internet: www.nss.org
Members: 25,000

> *the National Space Society (NSS) is a powerful advocate of space exploration and development*

With more than 25,000 members in thirty-five countries, and 95 chapters in the United States, Canada, Mexico, Australia, Ireland, and the United Kingdom, the National Space Society (NSS) is a powerful advocate of space exploration and development. Formed in 1987 from the merger of the L5 Society and the National Space Institute, the organization's stated mission is to promote change in social, technical, economic and political conditions to advance the day when people will live and work in space. A daunting mission, but one the NSS pursues with vigor and with the knowledge that both science and the ingenuity of humanity are on their side. The Society maintains an active program of education through papers, forums and symposia to legislators in Washington, DC. NSS Chairman of the Board of Directors is Apollo 11 astronaut, Buzz Aldrin. NSS Executive Director is David Brandt.

The National Space Society publishes *AD ASTRA* ("to the stars"), an attractive and informative non-technical bi-monthly magazine for members which reports on a wide range of space-related topics. The Society also holds an annual International Space Development Conference attended by more than 750 space activists.

Members of NSS are eligible to participate in Society-organized field trips to launch events around the country, including launches of the Space Shuttle, commercial ELV, and X-rocket. Specially-guided tours of NASA facilities, launch sites, and other facilities are also offered (as available).

For a list of the National Space Society chapters in the United States, Canada, Mexico, Australia, Ireland and the United Kingdom, see Appendix B.

Dues: $35 annual (individual)
$20 annual (seniors/students)

The Neville Public Museum Astronomical Society

210 Museum Place
Green Bay, WI 54304
Phone: (414) 448-4460
E-mail: ddewitt@online.dct.com
Members: 70

Meetings of the Neville Public Museum Astronomical Society are held the second Wednesday of each month from 7 - 9 p.m. at Neville Public Museum. Club meetings are informative and designed to educate new members in astronomy basics while continuing to educate more experienced members. Activities include an annual Astronomy Day and Public Observing Night, Messier Marathon, a Christmas Party, a summer picnic and observing weekends at Parmentier Observatory, a privately owned observatory with a 30-inch classical Cassegrain telescope. The Society publishes *The Eyepiece*, a monthly newsletter for members and is a member society of The Astronomical League. A 10-inch Dobsonian style club telescope is available for member use.

Dues: $15 annual

New Jersey Astronomical Association, Inc.

P.O. Box 214
High Bridge, NJ 08829-0214
Phone: (908) 638-8500
E-mail: kevtk@aol.com
Internet: www.njaa.org
Members: 200

In the winter of 1965, seven men with a vision, various talents and great determination formed the New Jersey Astronomical Association (NJAA). They wanted to build a large observatory for amateur astronomers and, to this end, they scoured the hills and towns for the only thing they needed to make their dream a reality - money. Many volunteers helped them finance this ambitious project and today, The Paul Robinson Observatory and the Edwin E. Aldrin, Jr. Astronomical Center stand on 18 acres at an elevation of almost 900 feet in Voorhees State Park in beautiful Hunterdon County.

The observatory houses a 26-inch, f/4 Newtonian reflector, the largest telescope available to the public in the state. The observatory is open to the public each Saturday evening and Sunday afternoon from June through October, and the fourth Saturday evening of the month the rest of the year. The Hunterdon County Adult School also offers courses in astronomy at the observatory.

The club is involved in many facets of astronomy. Meteor observing for the International Meteor Organization (IMO) and the North American Meteor Network (NAMN) are done on the grounds. Supernova and asteroid searches are also done with the help of the group's CCD camera. Conventional astrophotography workshops are given throughout the year. The NJAA also has a number of "Challenge Lists" to hone ones observing skills. The lists include searches for asteroids, lunar objects, and various deep sky objects. Completion of a "list" earns that member a spot on a special plaque denoting the accomplishment. Other activities sponsored by the club include an annual Messier Marathon in the spring and a similar "Fall Fling" in the autumn.

Membership benefits include a monthly newsletter, discounts in the club gift shop, reduced rates for various astronomy magazines, access to viewing yards for personal observing sessions, and access to the NJAA's own astroweather hotline for weather information in planning those sessions. Membership also entitles a member to take the Qualified Observer Course. This course instructs the member on how to use the 26-inch telescope for their own personal observing in exchange for a few "Public Duty Nights" a year.

Dues: $25 annual
        $40 annual (Sustaining)

The Association's Paul Robinson Observatory and the Edwin E. Aldrin, Jr. Astronomical Center stand on 18 acres at an elevation of almost 900 feet in Voorhees State Park

North American Meteor Network

7304 Doar Road
Awendaw, SC 29429
E-mail: MeteorObs@charleston.net
Internet: medicine.wustl.edu/
~kronkg/namn.html

founded . . . to coordinate the observations of amateur meteor observers

The North American Meteor Network (NAMN) was founded by Mark Davis in June 1995 to coordinate the observations of amateur meteor observers. The program has expanded to include recruiting new observers and providing guidance and instructions on how to conduct meteor observations that insure the collection of scientifically useful data. A newsletter is published monthly and is available to members free of charge.

The NAMN supports the work of both the Meteors Section of the Association of Lunar and Planetary Observers and the International Meteor Organization. All collected data is shared with these partner organizations, providing NAMN members with the widest possible dissemination and making member contributions available worldwide.

Dues: None

North East Wisconsin Stargazers (NEWSTAR)

514 Union Ave Apt E
Oshkosh, WI 54901-4385
Phone: (414) 426-2286
E-mail: tlb@vbe.com
Members: 50

North Jersey Astronomical Group

P.O. Box 1472
Clifton, NJ 07015-1472
Phone: (973) 614-9220
Internet: www.csam.montclair.edu/
~westk/njag.html
Members: 50

Curiosity and a sense of humor are the only real requirements for membership in the North Jersey Astronomical Group (NJAG), a member of the United Astronomy Clubs of New Jersey (UACNJ). Members are entitled to use the group's telescopic equipment and dark sky sites. Beginners are welcomed and encouraged and can learn from more experienced members. Meetings are held the second Wednesday of each month at Montclair State University (MSU). Members receive *Dark Sky Observer*, NJAG's monthly newsletter.

Dues: $15 annual (individual)
        $10 annual (MSU students)
        $20 annual (family)

Northeast Nebraska Astronomy Club

710 South 4th Street
Norfolk, NE 68701
E-Mail: gadams@kdsi.net (Gregg
Adams, Secretary/Treasure)
Members: 11

The goal of the Northeast Nebraska Astronomy Club is to provide education and viewing opportunities for area youth, to provide the general public with helpful astronomical information, and to share in dark sky observing of deep sky objects with other club members on a regular basis.

Dues: $15 annual

Northern Colorado Astronomical Society

7348 Poudre Canyon Hwy.
Bellvue, CO 80512
Phone: (970)493-7613
E-mail: bjarvis@ezlink.com
or: djlaszlo@aol.com,
rmoench@lamar.colostate.edu
Internet: lamar.colostate.edu/
~rmoench/ncasrdm.html
Members: 80

Members in the Northern Colorado Astronomical Society come from both Larimer and Weld counties in northern Colorado. The club meets at 7 p.m. on the first Thursday of each month at the Discovery Center Science Museum in Fort Collins. Monthly star parties are held at either the Pawnee National Grasslands near Ault, Colorado (in the winter), or at Deadman Mountain, near Red Feather Lakes Colorado at an elevation of 10,000 feet (in the summer). The club sponsors or assists with a number of events and programs, including Astronomy Day, public star parties, school events, Elderhostel programs and additional monthly star parties at Rocky Mountain National Park during the summer.

Membership benefits include a monthly newsletter and use of the club's 10-inch Dobsonian telescope. Members also receive discounts on magazine subscriptions.

Dues: $15 annual

Northern Kentucky Astronomers

Eric T. Costello, Director
2395 South Main Street
Highland Heights, KY 41076
Phone: (606) 441-4359
Members: 15

Reforming after several years absence, the Northern Kentucky Astronomers (NKA) has sites in Kentucky and Indiana for member use. Telescopes range from 4.5 to 13.1 inch, mostly Newtonian and Dobsonian. Several meetings each year and some star party sessions are open to the public. The club has placed a high priority on meteorite observations.

Dues: $15 annual

Northern Virginia Astronomy Club

5554 Sequoia Farms Dr.
Centreville, VA 20120
Phone: (703) 803-3153
E-mail: pjohnson@dgsys.com
Internet: astro.gmu.edu/~novac
Members: 300

The Northern Virginia Astronomy Club (NOVAC) meets the third Wednesday of each month at the Arlington Planetarium. Club meetings are open to the public. Meetings usually include guest speakers and sky tours using the planetarium facility. Membership benefits include discounts on *Astronomy* and *Sky & Telescope* magazine subscriptions, discounts on books, access to the club astronomy library, and use of club loaner telescopes. Additionally, NOVAC has several dark sky sites throughout Northern Virginia where members can gather on scheduled club nights.

Ohio Valley Astronomical Society

1827 Enslow Blvd.
Huntington, WV 25701
Phone: (304) 525-3849
E-mail: ball3@marshall.edu
Internet: webpages.marshall.edu/
~ball3/ovas.html
Members: 8

The Ohio Valley Astronomical Society (OVAS) meets the first Saturday of each month at 7:30 p.m. at the Union Missionary Baptist Church in Bradrick, Ohio. The Society is open to anyone with an interest in astronomy, regardless of background. Telescopes owned by members include a 6-inch, an 8-inch, and a 10" Newtonian, plus the Society has access to a 14-inch Schmitt-Cassegrain telescope in an observatory.

Dues: $12 annual (individual)
$18 annual (family)

Oklahoma City Astronomy Club

c/o Omniplex Science Museum
2100 N.E. 52nd Street
Oklahoma City, OK 73111
Phone: (405) 424-5545

Oklahoma City Astronomy Club is an active group of amateurs who offer many public events throughout the year, including the Okie-Tex Star Party which is attended by about 300 amateur astronomers annually. The club meets at 7 p.m. on the second Friday of each month. Guests are welcome.

Dues: $20 annual (family)

Olympic Astronomical Society

P.O. Box 458
Keyport, WA 98345
Members: 80

Omaha Astronomical Society

c/o Bill O'Donnell
4509 Seward
Omaha, NE 68104-5139
Phone: (402) 553-3448
E-mail: strgzr@ne.uswest.net
Internet: www.top.net/cdcheney
Members: 80

The Omaha Astronomical Society (OAS) has a membership that represents all areas of astronomical expertise. They are active in the promotion of astronomy education and public awareness and work regularly with the Neale Woods Nature Center in this capacity. The OAS meets on the first Friday of each month in room 170 of the Durham Science Center on the University of Nebraska, Omaha campus. The club also works with The Prairie Astronomy Club on such activities as the annual "Nebraska Star Party" at Meritt Reservoir, near Valentine, Nebraska, and public star parties at Mahoney State Park. The OAS has a 4-acre dark-sky site located in Weeping Water, Nebraska. The site is home to several observatory buildings, including a home for the club's 4-inch refractor.

Membership benefits include a subscription to the club's monthly newsletter, *The STELLA*, membership in the Astronomical League, a telescope loaner program, and access to the Weeping Water site.

Dues: $25 annual (regular/family)
$10 annual (junior/student)
$10 annual (newsletter only)

Orange County Astronomers

P.O. Box 1762
Costa Mesa, CA 92628
Phone: (714) 722-7900
Starline: (714) 995-2203
E-mail: JRSanf@aol.com
Internet: www.chapman.edu/oca/
Members: 600+

One of the largest astronomy clubs in the nation, the Orange County Astronomers (OCA) was founded in 1967 and incorporated in 1972. The club meets for a social-educational evening on the second Friday of each month, starting at 7:30 p.m. at the Irvine (formerly Hashinger) Science Hall of Chapman University. Star parties are held monthly at sites maintained by the OCA. The main star party is held at the Anza Observatory and Site, a 20-acre parcel owned and developed by the club and featuring large drive-in pads, an RV parking area, 46 concrete pads leased yearly to members for setting up telescopes, and the club's 22-inch (55cm) aperture Cassegrain William Kuhn Telescope. The Anza Star Party is traditionally held the Saturday night nearest the new moon. The OCA also leases a small observing site near the intersection of Santiago Canyon Road and Silverado Canyon Road in the foothills of Orange County. This site has a few telescope pads and piers and a rented toilet. Orange County Astronomers star parties go on from dusk till dawn, so they ask that you be careful of car lights near and on the site.

Membership benefits include a telescope loaner program (from small portable units up to a 16-inch Newtonian), a subscription to the club's 12-page monthly newsletter, the *Sirius Astronomer*, discounts on magazine subscriptions, good deals on used telescopes, and more. For those living more than 100 miles from Orange County, a special class of membership called *CyberMember* is available. This class of membership includes a subscription to the *Sirius Astronomer*, space on the web site, access to the club image library, and videotapes of meetings.

Dues: $35 annual (Regular)
       $20 annual (CyberMember)

The Pajarito Astronomers

P.O. Box 1092
Los Alamos NM 87544
Phone: (505) 662-3041
E-mail: daveh@lanl.gov
Members: 50

The Pajarito Astronomers is an organization of 50 professional and amateur astronomers that sponsors lectures for the public 6 times a year and provides dark-sky observing sessions for the public 8 times a year. The club is also involved in a number of public education activities throughout northern New Mexico each year. Club members receive a monthly newsletter and monthly sky calendar.

Dues: $7 annual

The Patient Astronomer, Ltd.

John T. Stewart, Director
2711 Harmon Road
Silver Spring, MD 20902
Phone: (301) 933-1596

The primary goal of The Patient Astronomer, Ltd. (PAL) is to expose the public to basic astronomy through direct experiences such as naked-eye and telescopic observations of the night sky. The organization publishes *The Celestial Watch*, a monthly newsletter of PAL community activities and upcoming events in the local sky. Members and non-profit organizations can arrange for The Patient Astronomer, Ltd. to hold a sky party. PAL will provide sky charts, an 8-inch motorized telescope, and a brief orientation talk using the sky charts. Slide shows, videos, demonstrations, and hands-on activities can also be arranged. Interested individuals can become either a member or a contributor (with the latter receiving special recognition and the knowledge that their support helps to bring astronomy to many who might not otherwise have such an opportunity).

Dues: $15 annual
       $50+ (Contributors)

Peninsula Astronomical Society

P.O. Box 4542
Mountain View, CA 94040
Phone: (415) 949-7334
E-mail: pasinfo@usa.net
World Wide Web: www.skyglass.com/pas.html
Members: 250

The Peninsula Astronomical Society is comprised of members from in and around the San Francisco Bay area representing all levels of interest, from rank beginner to the professional level. The Society meets monthly at Foothill College, in Los Altos Hills, on the second Friday of the month. Meetings are held in room F1 of the Forum building at 7:30 p.m., and all are welcome. Parking at Foothill College requires a permit which can be purchased at machines in the lots for $2 (in bills).

The Society runs the observatory at Foothill College and members may help out with the public observing programs there. The observatory is equipped with a 6" refractor and a 16" Newtonian reflector. It is open to the public on clear Friday evenings from 8:30 p.m. to 11:00 p.m. The observatory is also open on clear Saturday mornings from 10:00 am until Noon for solar viewing, in both h-alpha and white light.

In addition, the club has a private facility in the Santa Cruz mountains (at an elevation of 2400 feet) above the lights and smog of Silicon Valley. Here they have a 12" Tinsley Cassegrain, a 16" Newtonian, and a 28" Cassegrain (work in progress), along with terraces and electrical power for members to set up their own telescopes.

Numerious public observing programs are offered by the Society each year at locations throughout the Bay Area. One such program is the annual Yosemite Public Star Party, held at Glacier Point in Yosemite National Park.

Dues: $20 annual
$6 annual (under 18)

Phoenix Astronomical Society

Mail:
Terri Finch,
PAStimes Newsletter Editor
10828 N. Biltmore Drive #141
Phoenix, AZ 85029
Phone: (602) 547-2420
E-mail: starstuf@aztec.asu.edu
Members: 50

The members of the Phoenix Astronomical Society have many related interests, among these are telescopes, binoculars, astrophotography, computer planetaria, astronautics, skylore, and celestial navigation. Established in 1948, the Society includes both amateur and professional astronomers and meets once each month, usually on the first Thursday of the month at 7 p.m., at the Brophy College Preparatory Physics Lab (4701 N. Central Avenue). Bi-monthly outings to a desert site north of Phoenix are also held, offering members the opportunity to use telescopes, learn their way around the night sky, and share observing techniques. The Society sponsors educational presentations for area schools and youth groups. Membership benefits include a subscription to *PAStimes*, the Society's monthly newsletter, access to the club library (books, periodicals, maps, slides, and tapes), and access to the club telescope, an 8-inch Newtonian. Anyone can be placed on the Society's mailing list and receive the *PAStimes*

newsletter free for two months by contacting Terri (Renner) Finch at the above address.

Dues: $21 annual
      $48 annual (includes *Sky & Telescope* subscription)

The Planetary Society
65 North Catalina Avenue
Pasadena, CA 91106-2301
Phone: (626) 793-5100
Fax: (626) 793-5528
E-mail: tps@mars.planetary.org
Internet: planetary.org
Members: 100,000+

With 100,000 members in over 100 countries, the Society is the largest space interest group on Earth

The Planetary Society was founded in 1980 by Drs. Carl Sagan, Bruce Murray and Louis Friedman to encourage the exploration of our solar system and the search for extraterrestrial life. The Society is a nonprofit, non-governmental organization, funded by dues and donations from individuals around the world. With 100,000 members in over 100 countries, the Society is the largest space interest group on Earth. Membership is open to all people interested in their mission. The mission of the Society is to encourage all space faring nations to explore other worlds, provide public information and support educational activities about the exploration of the solar system and the search for extraterrestrial life, and support and fund innovative and novel research and development projects that can seed future projects of planetary exploration.

Membership benefits include a full year's subscription to the Society's acclaimed bimonthly magazine, *The Planetary Report*, an invitation to subscribe to the Society's other distinguished Members-only publications, an official Planetary Society Membership Card, identifying you as a Member in good standing and admitting you to Society-sponsored events and activities, discounts on a variety of space-related books, prints, slides, posters and other educational materials (many are Society exclusives!), and invitations to Society-sponsored conferences, seminars, lectures and films, workshops and other events.

The Planetary Society's store offers a wide variety of books, gift ideas, and other space-related items. Current and back issue copies of *The Planetary Report*, *Mars Underground News*, *Bioastronomy News*, and *NEO News* are also available.

Pontchartrain Astronomy Society, Inc.

Mike Sandras, Planetarium Director
Daily Living Science Center
409 Williams Blvd.
Kenner, LA 70062
Phone: (504) 468-7229
Members: 150

The Pontchartrain Astronomy Society (PAS) conducts a General Meeting (open to the public) once a month to discuss club business, view presentations, and socialize. An Observers Group meeting is also held each month for the more serious observers. Prospective members don't need to own a telescope or have knowledge of astronomy. Member benefits include a subscription to *PAS-Times*, the clubs monthly newsletter, a subscription to the *Reflector*, a quarterly publication from the Astronomical League, discounts on astronomy magazines, free membership in the Astronomical League, an annual picnic, free admission to the Planetarium and Observatory in Kenner, LA, and access to the club library.

Dues: $20 annual

Popular Astronomy Club, Inc.

John Deere Planetarium
Augustana College
Mail:
2535 45th Street
Rock Island, IL 61201
Phone: (309) 786-6119
E-mail: prc29@aol.com
Members: 70

Founded in 1936, the Popular Astronomy Club (PAC) meets on the second Monday of each month, except August, when the meeting is held on the night of the Perseid Meteor Shower, and October, when they schedule an Annual Banquet. Meetings begin at 7:30 p.m. The Popular Astronomy Club is a member of the Astronomical League and members receive the newsletter of the Astronomical League in addition to the PAC newsletter.

Dues: $12 annual (general)
$5 annual (student)

Rockland Astronomy Club of New York

73 Haring Street
Closter, NJ 07624-1709
Phone: (201) 768-6575
Message Center: (914) 786-7730
E-mail: durban1@aol.com
Internet: members.aol.com/
StarshipCS/RAC/index.htm
Members: 180 (20-30 active)

The Rockland Astronomy Club of New York holds monthly public star parties on Saturday nights at state parks in New York State from April to October. The club is active in the community and sponsors the Northeast Astronomy Forum & Telescope Show each April (most recent attendance exceeded 1,000 people), and the Summer Star Party each July or August at Shady Pines Campground in Savoy, Massachusetts.

Royal Cabrillo Astronomical Stargazing Society

Cabrillo College, Astronomy Dept.
6500 Soquel Drive
Aptos, CA 95003
Phone: (408) 864-8282
E-mail: vonahnen@admin.fhda.edu

The Royal Cabrillo Astronomical Stargazing Society holds public star parties every month, usually on the Saturday evening closest to the new moon (weather permitting).

Royal Greenwich Observatory

Madingley Road
Cambridge CB3 0EZ
United Kingdom
Phone: 44-1223-374000
Fax: 44-1223-374700
E-mail: mjp@ast.cam.ac.uk
Internet: www.ast.cam.ac.uk/RGO

founded by Charles II in 1675 at Greenwich

The Royal Greenwich Observatory (RGO) was founded by Charles II in 1675 at Greenwich. It was transferred to Herstmonceux Castle, Sussex, England after World War II and is now a facility of the Particle Physics and Astronomy Research Council (PPARC). Past directors include such prominent astronomers as Edmond Halley and James Bradley.

Ryerson Astronomical Society

1100 E. 58th St #8
Chicago, IL 60637
E-mail: astro-request
@lists.uchicago.edu
Internet: astro.uchicago.edu/home/
web/RAS/

The purpose of the Ryerson Astronomical Society is to encourage all humanity to "behold the celestial lumi-naries in hushed awe." The Society is a registered student organization at the University of Chicago and has existed since at least 1938. Carl Sagan was a member as a student. The Society's observatory is located on top of Ryerson Physical Laboratory.

Sacramento Valley Astronomical Society

P.O. Box 575
Rocklin, CA 95677-0575
Phone: (916) SVAS-111
Internet: www.calweb.com/~SVAS
Members: 220

Established in 1945, the Sacramento Valley Astronomical Society (SVAS) owns and operates the Henry Grieb Observatory, located at Blue Canyon Airport (5,300' elevation) in the Sierra Mountains. Meetings are held the third Saturday of each month at the Discovery Museum and Learning Center (3615 Auburn Boulevard, Sacramento).

Dues: $25 annual (General)
       $75 annual (Observatory
       Members)

San Diego Astronomy Association

P.O. Box 23215
San Diego, CA 92193
Office: (619) 645-8940
Observatory: (619) 766-9911
E-mail: sdaa@digcir.cts.com
E-mail Membership Requests:
sdaainfo@digcir.cts.com
Internet: www.geocities.com/
CapeCanaveral/2419/sdaa_hom.htm
Members: 375

Established in 1963, the San Diego Astronomy Association (SDAA) is a non-profit education corporation established in 1963. The SDAA's purpose is to further the education of its members and the general public in the subjects of astronomy and the related space and physical sciences, and to increase public awareness and enjoyment of these sciences.

The membership is composed of people from all walks of life who share a common scientific interest. Many have built their own telescopes. Some are involved in school and civic activities. Some conduct research for professional organizations, while others just enjoy the serenity of a night under a canopy of stars. Whatever a member's interest, all share a common fascination for the mysteries of our universe and a willingness to exchange and share the results of practical observational experiences.

The SDAA holds monthly meetings the third Friday of each month, starting at 7:30 p.m., in the Grayson Boehm Hall of the Reuben H. Fleet Space Theater and Science Center, Balboa Park, San Diego. Speakers come from JPL, Mt. Palomar Observatory, local universities, and commercial manufacturers, among others. Preceding the meeting there is a Beginner's Program starting at 7:00 p.m., aimed at getting the novice further involved in astronomy. On the first Wednesday of the month, SDAA members set up telescopes outside the Fleet Space Theater for public observing. All meetings are open to the general public, and admission to all SDAA events is free.

The SDAA hosts approximately sixty public star parties a year, with 75% of them for the public school system. Members volunteer their time, equipment, and love of the skies to share the heavens with approximately 10,000 students and adults each year. The SDAA has a ten-acre dark-sky observing site located 75 miles east of San Diego. The site has public and private observing pads complete with electricity, bathroom and shower facilities, and an observatory housing the Lipp 22" Telescope, an f/7.4 Ritchey-Cretian design. The observatory and site are open two nights a month for public use and enjoyment. Contributing members may use the site at any time, and be trained on the club telescope for personal use.

Other benefits of membership include a monthly newsletter, an extensive print, video, and software library, an online service provided by Digital Circus BBS ((619) 233-5348), mirror grinding classes, field trips to world renowned sites of astronomical interest (i.e., NASA/JPL, Mt. Palomar Observatory, Mt. Wilson Observatory, The Mexican National Observatory in Baja, Mexico, and Kitt Peak), an annual banquet with camaraderie and numerous door prizes, and summer and fall barbecues. Members also receive discounts from Sky Publishing, The Astronomical Society of the Pacific, and from various equipment vendors. The SDAA is committed to bringing the Universe within the reach of anyone interested.

Dues: $30 annual (Contributing)
        $22 annual (Basic)

San Jose
Astronomical Association

3509 Calico Avenue
San Jose, CA 95124-2536
Phone: (408) 559-1221
E-mail: Jim.Van.Nuland@sjpc.org
Internet: www.seds.org/billa/sjaa/
sjaa.html
Members: 200

Affiliated with the Astronomical League, Astronomical Association of Northern California, and Western Amateur Astronomers, the San Jose Astronomical Association is a busy club with events on most weekends and (during the school year) on some weekday evenings. The association is oriented towards observing and their calendar is irregular in order to maximize the number of dark-sky star parties. All activities are open to the public, with the only cost being admission to the state park.

The club holds a general meeting monthly, normally on the Saturday nearest the full moon. In April or May the association holds an Astronomical Swap Meet and Auction, which replaces that month's general meeting. A Beginner's Observational Astronomy class is offered by the club ten times a year. There is no fee or required purchases for this class and membership is not required. Normally, two public star parties are held each month in a city park. Additionally, one or two deep-sky star parties are held on Saturdays near the new moon at Henry Coe State Park or Fremont Peak State Park. At the invitation of schools, about 30 star parties are held each year, mostly on weekday evenings, at school yards.

Membership benefits include a telescope loaner program, subscription to the club's monthly newsletter, coverage by group liability insurance, and discounts on magazine subscriptions, books, charts, and other items.

Dues: $15 annual

San Mateo
Astronomical Society

P.O. Box 974, Station A
San Mateo, CA 94403
Phone: (415) 594-9686
Members: 80

Founded in 1960, the San Mateo Astronomical Society meets the first Thursday of each month (except July and August) and usually holds star parties on the Saturday before the new moon. The club publishes a monthly bulletin, which is available to members (a complimentary bulletin is available when requested in writing).

Seattle Astronomical Society

P.O. Box 31746
Seattle, WA 98103
Information: (206)-523-ASTR
E-mail: xx004@scn.org
**Internet: www.scn.org/ip/sastro/
sas.html**
Members: 225

An annual Astronomy Day display is put on by the SAS in conjunction with the Pacific Science Center

The Seattle Astronomical Society (SAS) serves amateur astronomers in the Seattle metropolitan area and western Washington state. Meetings are held on the third Wednesday of each month at 7:30 p.m. in a lecture hall at the Astronomy-Physics Building, University of Washington campus. Meetings are open to the public.

The SAS holds a public star party on the Saturday closest to each first quarter moon, at Green Lake Park. Other star parties are held by the Society's Sidewalk Astronomers, Lunar Observers group, and Dark Sky Observers group, as weather permits. There are also SAS groups for Telescope Makers, Astrophotography, and

Computers in Astronomy. SAS publishes a monthly newsletter, *The Webfooted Astronomer*.

An annual Astronomy Day display is put on by the SAS in conjunction with the Pacific Science Center. The Society also responds to many requests from school groups to provide astronomy information, telescope demonstrations and hands-on viewing sessions. For special events such as eclipses and comets, the SAS organizes city-wide public viewing sessions at multiple sites.

Public information about the SAS and astronomical events is available on the Society's telephone Information Line, on their WWW page, or via e-mail inquiry. In addition, members have access to a Dark Sky Observers voice mail service, and to a computer e-mail list called webftweb. Other membership benefits include optional discount subscriptions to *Sky & Telescope* or *Astronomy* magazine, membership in the Astronomical League, and access to several club telescopes which are available for loan.

Dues: $25 annual (general)
         $10 annual (student)

The SETI League, Inc.

433 Liberty Street
Mail:
P.O. Box 555
Little Ferry NJ 07643
Phone: (201) 641-1770
Fax: (201) 641-1771
E-mail: join@setileague.org
Internet: www.setileague.org/
Members: 1000 (in 30 countries)

The SETI League, Inc. is a membership-supported, educational and scientific corporation, dedicated to the electromagnetic Search for Extra-Terrestrial Intelligence. Members are helping to continue the research Congress wouldn't let NASA finish. The League's Project Argus search of the heavens for coherent radio signals of intelligent extra-terrestrial origin was launched in April 1996. When fully implemented, it will ultimately involve 5,000 small radio telescopes, built and operated by amateurs, coordinated to view the entire sky in real time.

The primary mission of the SETI League is to encourage and support the Search for Extra-Terrestrial Intelligence by attracting interested radio amateurs, astronomy enthusiasts, and microwave and digital signal processing experimenters into the SETI community; by developing technologies to assist the advanced experimenter in assembling a workable SETI receiving station; by disseminating hardware and software designs in support of SETI; by coordinating SETI experimenters worldwide in conducting a thorough sky survey; by providing a medium for communication between SETI experimenters, enthusiasts, and organizations, through journals, meetings, conferences, and electronic means; by assembling, maintaining and operating advanced optical and radio telescopes; by identifying and publicizing potential spin-off applications of SETI technologies; by encouraging the restoration of public funding for the NASA SETI Office; and by raising public consciousness as to the importance and significance of a broad-based Search for Extra-Terrestrial Intelligence.

Membership benefits include a subscription to *SearchLites*, a quarterly newsletter, use of the SETI research library, access to technical support, books, manuals, conferences, meetings, e-mail discussion groups, memorabilia items (T-shirts, coffee cups, mouse pads, etc.), and the opportunity to help discover humanity's place in the cosmos.

Dues: $50 annual (Full Membership)
7 other membership categories are available

Sheboygan Astronomical Society

Diane Van Minsel, President
3728 Mary St.
Manitowoc, WI 54220-5458
Phone: (414) 686-9234
Fax: (414) 686-9234
E-mail: dvminsel@lakefield.net
Internet: bratshb.uwc.edu/~sas
Members: 23

very active in the "Rockets for Schools" program, which began in 1996

The Sheboygan Astronomical Society holds monthly meetings at the U.W. Sheboygan center on the first Tuesday of each month, September through May. Meetings begin at 7 p.m. and are open to the public. Each month there is a general topic with plenty of discussion time at the end of the presentation. Outside speakers such as Robert Naeye from *Astronomy* magazine and John Briggs of Yerkes Observatory have been past guests. The Society publishes a monthly newsletter and they have a web page that includes a chat room for discussing astronomical topics. The web page is updated monthly.

The Society sponsors local festivities in celebration of National Astronomy Day. Items which have proven to be very popular during these events include solar observing and the portable planetarium, which allows the general public to see the night sky in the daytime. The Sheboygan Astronomical Society has been awarded Honorable Mentions from the National Astronomical League two years running for these efforts.

The Society does school programs and programs for the Boy Scouts and Girl Scouts and is very active in the "Rockets for Schools" program, which began in 1996. Two rockets will be launched in 1997 and Society members have constructed an antenna which will receive signals from the rockets. This antenna will also receive signals from the Mars probes as they approach and land on Mars.

Dues: $10 annual

The Shreveport-Bossier Astronomical Society, Inc.

Jennie Goodwin, Secretary/Treasurer
353 Ockley Drive
Shreveport, LA 71105-2917
Phone: (318) 865-2433
Internet: www.lsu.edu/nonprofit/
sbas/
Members: 60

Founded in 1959, the Shreveport-Bossier Astronomical Society is a member of the Astronomical League and holds meetings on the third Saturday of each month at 7:30 p.m. at the Worley Memorial Observatory south of Shreveport. Star parties are held for the public throughout the year. Membership is open to anyone 10 years of age or older (contact the Secretary for info on family rates). Members receive a monthly newsletter.

Dues: $17 annual (individual)

Sidewalk Astronomers

1946 Vedanta Pl.
Hollywood, CA 90068
Phone: (818) 841-0548
E-mail: bagheer@worldnet.att.net,
atmavidya@vedanta.org,
bigguy1@pacbell.net
Internet: www.geocities.com/
CapeCanaveral/6389

Inspired by John Dobson, the former monk who popularized the Dobsonian telescope

Inspired by John Dobson, the former monk who popularized the Dobsonian telescope mount and helped bring the stars and planets to countless people, the San Francisco Sidewalk Astronomers was formed in 1968. Dobson, who was expelled from his monastery in 1967 because of his telescope-making (nefarious activity that it is), began setting up his unique telescopes at the corner of Broderick and Jackson in San Francisco on clear nights. Passers-by would peer through the telescopes while John explained what they were seeing. From there, the organization grew, spreading across the country. Eventually "San Francisco" was dropped from the name and the organization is now referred to as simply The Sidewalk Astronomers.

The organizations roots play an important part in their philosophy and they still gather with telescopes at public sites, such as street corners, malls, parks, camps, schools, and libraries, allowing passers-by to view the heavens. The Sidewalk Astronomers is also committed to keeping the art of telescope-making and telescope mirror-making alive. As part of this commitment, telescope-making classes are offered by members in several areas around the country. You might even get some pointers from Mr. Dobson himself. The organization responds to a steady stream of letters from individuals, schools, and other entities around the country who need help with telescope or mirror-making, or need sources for materials.

For those who cannot attend a telescope-making class sponsored by The Sidewalk Astronomers, a 90-minute telescope-making video is now available featuring Mr. Dobson, who takes you through every step of the telescope-making process, from making the mirror to building the mount.

The video is available from Dobson Astro Initiatives, P.O. Box 460915, San Francisco, CA 94146-0915. The price is $39.95 plus $3.50 shipping and handling (California residents add $3.40 sales tax). A free color flier is available upon request.

Skokie Valley Astronomers

Gretchen Patti
1 S. 751 Avon Drive
Warrenville, IL 60555
Phone: (630) 393-7929
Fax: (630) 505-9292
E-mail: gretchen@adichi.com
Internet: ourworld.compuserve.com/
homepages/E_Neuzil
Members: 25

This collection of amateur enthusiasts includes school kids, retirees, telescope makers, teachers, armchair astronomers, astrophotographers, and even a couple of folks with astronomy degrees. Meetings are held in Deerfield on the second Friday of each month at 8 p.m. at the Ryerson Nature Center. The indoor presentation is followed by observing (weather permitting) using members' telescopes. All meetings are open to the public.

Dues: $20 annual (family)

The Skyland Stargazers

100 Tanglewood Drive
East Hanover, NJ 07936
Phone: (201) 887-5866
Fax: (201) 386-0969
E-mail: apisano@ix.netcom.com
Members: 12

The Skyland Stargazers was founded in 1989 and is the sister club to the Morris Museum Astronomical Society. The club meets on the third Friday of every month and holds astronomical observing sessions throughout the year. The Skyland Stargazers have their own observatory which houses an 18-inch Newtonian telescope at Jenny Jump State Park in western New Jersey.

Dues: $15 annual

Skyscrapers, Inc.

Seagrave Observatory
47 Peeptoad Road
North Scituate, RI
Internet: chandra.cis.brown.edu/
astro/skyscrapers
Members: 90

Skyscrapers was founded in 1932 by the late Dr. Charles Smiley, of Brown University, as the Amateur Astronomical Society of Rhode Island. The Society incorporated in 1936 and purchased Seagrave Memorial Observatory in North Scituate, Rhode Island from the estate of Frank Evans Seagrave, a noted amateur astronomer. Members include knowledgeable amateur astronomers and beginning observers who get together regularly to discuss and learn about the science of astronomy.

Skyscrapers' Seagrave Memorial Observatory features a variety of telescopes, including an 8-inch Alvan Clark refractor, a 12-inch Patton reflector, a 12-inch Meade computer controlled Schmidt-Cassegrain telescope, and 10, 14, and 20-inch Dobsonian reflectors. The observatory grounds also house a meeting hall which is used for Society meetings. Seagrave Observatory is frequently open to the public and is available for visits by school classes, scout groups or any other interested groups.

Skyscrapers holds regular meetings, usually on the first Friday of each month at 8 p.m. at Seagrave Memorial Observatory. During the winter months meetings are held off-site when the grounds are inaccessible. Programs include topics that are of interest to the beginner and the experienced amateur, such as observing techniques, space exploration, astrophotography, telescopes, astrophysics, and planetary geology. Guests are always welcome. In addition to regular monthly meetings, Skyscrapers conducts star parties and workshops for members on a regular basis. Each fall, usually the first Saturday of October, Skyscrapers hosts Astro-Assembly. This event brings together amateur astronomers from all over the New England area and has been a tradition for over 35 years.

No special qualifications are necessary for membership in Skyscrapers. You need no knowledge of astronomy nor do you need to own a telescope. Many members have never owned a telescope of their own, preferring to use the fine instruments owned by Skyscrapers. Others have made their own telescopes with assistance and instruction from other members. All you need is an interest in the sky! Membership benefits include access to the club's reflecting telescopes and loan of the 10-inch reflector.

Skywatchers Astronomy Club

c/o M. E. Ogburn
124 Bowstring Drive
Williamsburg, VA 23185-4952
Phone: (757) 864-1175
Fax: (757) 864-7722
E-mail: m.e.ogburn@larc.nasa.gov
Members: 20

The club operates an observatory at NASA, which houses a 16-inch Newtonian telescope and several smaller telescopes

The Skywatchers Astronomy Club is based in Hampton, VA, and is affiliated with the NASA Langley Activities Association, although membership is open to anyone. The club operates an observatory at NASA, which houses a 16-inch Newtonian telescope and several smaller telescopes. In addition to events for club members, free observing sessions are held for the general public, and the club instructs student groups and Scout troops.

Meetings are held one Thursday each month at 7:30 p.m. A newsletter is distributed quarterly and members receive discounts on astronomical products and publications.

Dues: $20 annual

Society For Scientific Exploration

P.O. Box 3818, University Station
Charlottesville, Virginia 22903
Phone: (804) 924-4905
Fax: (804) 924-3104
E-mail: lwf@virginia.edu or
sims@jse.com
Internet: www.jse.com
Members: 640 members and
associates

The Society is open to serious discussion and investigation of any observation or theory which seems to contradict the accepted dogma. A recent meeting covered areas as diverse as cosmology experiments, dowsing experiments, zero-point field theories, reincarnation, biophotons, homeopathy, etc. Meetings are held annually in the U.S. and semiannually in Europe or the Pacific rim. Various study groups focus on given areas in detail and depth. Reports from the study groups are discussed at the annual meeting and disseminated widely.

Membership includes a subscription to the *Journal of Scientific Exploration* (quarterly) and *The Explorer* (quarterly newsletter). The *Journal of Scientific Exploration* is also available to nonmembers for $55 ($100 libraries).

Dues: $75 annual
$55 annual (associates)

Society of Amateur Radio Astronomers

Tom Crowley, President
3623 West 139th Street
Cleveland, OH 44111
Internet: wbs.net/sara.htm
Members: 400

If you have an interest in radio astronomy and would like to become an amateur radio astronomer, the Society of Amateur Radio Astronomers, known as "SARA," invites you to join. Amateur radio astronomy groups are relatively new. SARA was organized in 1981 and as of 1996 has approximately 400 members worldwide. SARA is the largest amateur radio astronomy group in the world, and is eager to welcome enthusiasts to their group. SARA's members are involved in Solar Radio Astronomy, Total Power Sky Surveys and Narrow Band SETI experiments. A yearly conference is normally held in July at the National Radio Astronomy Observatory located in Greenbank, West Virginia.

Radio astronomy is the observation of radio frequency emissions from space, versus the more conventional observation of visible light as seen through an optical telescope. The field of radio astronomy is approximately 60 years old, having started in 1932. A young physicist by the name of Karl G. Jansky. was working for Bell Labs in 1930, about the time trans-atlantic radio was in its prime using long waves. He was assigned the task of determining the source of radio static on the newer short wave frequencies, and if possible locate any specific sources of this interference. Mr. Jansky did not locate any specific directional source of the high level static interference but he did note that as he turned his antenna toward the galactic plane, (the Milky Way) he recorded an increase in static "hiss" in his receiver. After much testing and verification it was concluded that the radio hiss was coming from the center of the Milky Way. This revelation did not appear to create much discussion at the time and due to other work assignments, Mr. Jansky did not have the opportunity to pursue this discovery. Little did he realize that his work was to be the foundation of radio astronomy.

In 1937, a ham radio operator, Mr. Grote Reber in Wheaton, IL (W9GFZ), designed and built a 31.5 foot parabolic dish antenna to continue Mr. Jansky's work. At a receiving frequency of 137 MHZ Mr. Reber discovered discrete regions of intense radio emission or "hiss". Mr. Reber plotted the regions of the sky, noting the intensity of the radio sources. He produced a map of the sky showing visible star sources, the result of his effort was the start of radio astronomy. Although there may have been other professional investigations during this period of time, the works of Mr. Grote Reber, an amateur radio astronomer and a ham radio operator, is most inspiring to the members of SARA.

Those interested in joining SARA can send for a membership application. If you live outside of the U.S., SARA asks that you not send checks, but send postal money orders instead. It is also important to include a street address because post office boxes are not acceptable for all deliveries. Letters of interest and applications for membership should be sent to Hal Braschwitz at the address above.

Dues: $24 annual (U.S.)
$16 annual (U.S. 18 & under)
$30 annual (Canada - US$)
$36 annual (foreign - US$)

Southeastern Association for Research in Astronomy

Dept. of Physics & Space Science
Florida Tech
150 West University Blvd.
Melbourne, FL 32901-6988
Phone: (407) 674-8098
Fax: (407) 984-8461
E-mail: sara-info@pss.fit.edu

SARA operates a 0.9-meter Cassegrain design telescope at Mercedes Point on Kitt Peak

The Southeastern Association for Research in Astronomy (SARA) is a consortium of the Florida Institute of Technology, East Tennessee State University, the University of Georgia, Valdosta State University, and the Florida International University. Formed in 1989, SARA operates a 0.9-meter Cassegrain design telescope at Mercedes Point on Kitt Peak. This telescope was formerly located at the Kitt Peak National Observatory and was awarded to SARA after being decommissioned due to budget constraints. The Kitt Peak telescope is operated as a fully automated observing facility. Approximately 10% of each year's observing time is available to non-SARA astronomers on a contractual basis.

Southern Oregon Skywatchers

Jessica Vineyard, President
426 B Street
Ashland, OR 97520
Phone: (541) 482-0783
E-mail: galaxygl@mind.net
Members: 70

Southern Oregon Skywatchers was established in August 1993 in response to the growing interest in astronomy in the region. The Skywatchers want to make astronomy accessible to everyone in an informal, easy to understand format. Meetings are held on the third Wednesday of each month at the Medford Service Center, 821 N. Columbus, Medford, Oregon, starting at 7:30 p.m. The club also holds monthly star parties in dark sky locations, sponsors an outreach program and publishes *The Orbiter*, a monthly newsletter. Other benefits include membership in The Astronomical League, loaner telescopes, use of the club's astronomy library, and discounts to *Sky & Telescope* and *Astronomy* magazines.

Dues: $15 annual (individual)
          $20 annual (family)

Southwest Montana Astronomical Society

c/o Museum of the Rockies
600 W. Kagy Blvd.
Montana State University
Bozeman, MT 59717
Members: 50

The Southwest Montana Astronomical Society (SMAS) meets on the last Friday of each month (January through October) at 7:00 p.m. at the Museum of the Rockies. Meetings are always open to the public. Meetings may include a speaker, demonstration, workshop, telescope viewing, and/or reports of current club events. The club and the museum co-sponsor the annual Montana Astro Fair, an extravaganza of popular astronomy activities held at the Museum of the Rockies in late-Winter. SMAS itself sponsors the annual Montana Starwatch, a weekend camp out of dark sky observing in August that also includes guest speakers, a BBQ, door prizes and other activities. Benefits include a quarterly newsletter, *The Star Wrangler*, voting privileges, access to the club's "loaner scope," and discounts on popular astronomy magazines.

Space Frontier Foundation

16 First Ave.
Nyack, NY 10960
Phone: 1-800-78-SPACE
E-mail: openfrontier@delphi.com
Internet: www.space-frontier.org

dedicated to opening the Space Frontier to human settlement

The Space Frontier Foundation is an organization of people dedicated to opening the Space Frontier to human settlement as rapidly as possible. The Foundation's goals include protecting the Earth's fragile biosphere and creating a freer and more prosperous life for each generation by using the unlimited energy and material resources of space. By unleashing the power of free enterprise, the Foundation hopes to lead a united humanity permanently into the Solar System. To support their work, the Space Frontier Foundation requests that you send a donation of $25 ($35 outside the US).

Space Telescope Science Institute

Johns Hopkins Homewood Campus
3700 San Martin Drive
Baltimore, MD 21218

The Space Telescope Science Institute (ST ScI) conducts a long-term program of scientific exploration with NASA's Hubble Space Telescope (HST). Established in 1981, NASA selected the Association of Universities for Research in Astronomy to manage ST ScI as an astronomy and research center and international observatory. Today the ST ScI is successfully carrying out its primary mission of maximizing the scientific return from HST in cooperation with the international astronomical community.

The ST ScI provides long-term guidance for this ambitious scientific effort, engages the participation of astronomers around the world and disseminates the astronomical discoveries from HST to the scientific community as well as the public. The ST ScI's staff of 380 astronomers, computer scientists and technicians oversees the evaluation and selection of research proposals, carries out the complex tasks of planning and executing observations and processing daily the flood of HST data through an elaborate computer "pipeline."

The ST ScI contains 220,000 square feet of offices, work areas, laboratories, conference rooms, computer rooms, laboratories and a cafeteria for a full-time staff of 500. A total of 87 VAX computers are used for space telescope operations, system development and testing, and scientific research. ST ScI maintains a research library, high-volume photo processing laboratory, 188-seat auditorium with state-of-the-art audio-visual controls and displays for scientific colloquia and press conferences.

Note: The above information was extracted from the NASA web site and is gratefully acknowledged.

Springfield Stars Club

P.O. Box 2793
Springfield, MA 01101-2793
Phone: (413) 782-7054
E-mail: rsanderson@juno.com
Internet: www.reflector.org
Members: 70

The Springfield Stars Club was founded in 1934 by a group of local people who were interested in making telescopes. Since then, it has grown into a diversified organization of astronomy enthusiasts from all walks of life, with widely varied interests. Meetings are held at the Springfield Museum of Science on the fourth Tuesday of each month, from September to June, beginning at 7:30 p.m.

Astronomy popularization and public education have long been high priorities for the Stars Club, and the organization often works hand-in-hand with the Science Museum to accomplish these goals. Past projects include the construction of a Foucault pendulum and a 20-inch Schmidt-Cassegrain telescope and observatory, both as contributions to the Science Museum. The Stars Club conducts "Stars Over Springfield," a popular series of public viewing nights at the museum on the first Friday of each month, as well as numerous public outreach and school programs. The Stars Club has also established an astronomy hot-line at (413)263-6800 ext. 414.

Dues: $20 annual

Starfax Association

P.O. Box 6084
Asheville, NC 28816-6084
Phone: (704) 254-4470
Fax: (704) 254-0988
E-Mail: starfax@aol.com
Internet: pobox.com/~owl/
Members: 50

Serving the entire western North Carolina area, the Starfax Association meets once each month for a star party on Blue Ridge Parkway near Asheville. The Association is heavily involved in educational efforts with local scouting groups, civic organizations, and the general public.

Stillwater Stargazers

Brukner Nature Center
5995 Horseshoe Bend Rd.
Troy, OH 45373
Phone: (937) 698-2879
E-mail: garypike@bright.net
Internet: www.dma.org/~wagner

Stillwater Stargazers meet on the third Tuesday of the month at 7:30 p.m. at Brukner Nature Center. They also have a free public star gaze the third Saturday of the month at Brukner Nature Center. Call or write for details.

Stockton Astronomical Society

P.O. Box 243
Stockton, CA 95201
Members: 160

Founded 47 years ago, the Stockton Astronomical Society is open to anyone with an interest in the night sky. Meetings are held the second Wednesday of each month at the George H. Clever Planetarium at San Joaquin Delta College, Stockton, California. Members can also participate in two monthly star parties, one of which is a high-altitude overnight star party. A public viewing session is held monthly outside the planetarium to introduce as many people as possible, especially children, to the wonders of the night sky. Members receive *Valley Skies*, the Society's monthly newsletter.

Dues: $15 annual (family)
        $10 annual (student)

Students for the Exploration and Development of Space

MIT Room W20-445
77 Massachusetts Avenue
Cambridge, MA 02139-4307
Phone: (888) 321-7337
Fax: (617) 253-8897
E-mail: seds@seds.org
Internet: www.seds.org

> *SEDS is a chapter based organization with chapters throughout the United States, Canada, United Kingdom, Latin America, and the Middle East*

Students for the Exploration and Development of Space (SEDS) is an independent, student-based organization which promotes the exploration and development of space. SEDS pursues this mission by educating people about the benefits of space, by supporting a network of interested students, by providing an opportunity for members to develop their leadership skills, and inspiring people through our involvement in space-related projects. SEDS believes in a space-faring civilization and that focusing the enthusiasm of young people is the key to our future in space.

Students for the Exploration and Development of Space was founded in 1980 at MIT and Princeton and consists of an international group of high school, undergraduate, and graduate students from a diverse range of educational backgrounds who are working to promote space as a whole. SEDS is a chapter based organization with chapters throughout the United States, Canada, United Kingdom, Latin America, and the Middle East. The permanent National Headquarters for SEDS-USA resides at MIT. Each chapter is fairly independent and coordinates activities and projects in its own area.

SEDS members are people interested in doing as much as they can to promote space exploration and development. The first step in this lengthy process is learning. SEDS provides an excellent environment in which to obtain access to many sources of information including speakers, tours, films, discussion groups & daily NASA updates. Astronomical observing trips and tours of local space facilities also play a significant role in the life of many SEDS members. SEDS members often take the knowledge they have gained and use it to influence the future of the space program. Students at several chapters have played major roles in organizing large conferences and have established important contacts with members of the space community. Others have helped increase public awareness of the benefits of space exploration by offering presentations to local primary and secondary schools as well as universities. All chapters keep in contact with each other through on-line computer networks.

SEDS can provide an organized pathway for you to begin space-related projects and give you a chance to break away from the usual classwork. You can also use SEDS as a stepping stone to a space-related career. Being actively involved in SEDS can put you in touch with many members of the space, technology, and education community and will allow you to develop the experience necessary to take leadership roles in your future career, whatever that may be.

Tenn Valley Astronomy Club

Contact: T. L. Ingram
276 Swan Creek Road
Fayetteville, TN 37334
Phone: (615) 732-4777
Members: 53

The Tenn Valley Astronomy Club stands apart from most astronomy clubs because they are not involved with club business interests, budgets, building and library maintenance, politics, or anything else in astronomy except their one interest, observing. Because the club meets only to observe, there are no dues.

This is a good club for those wanting to get in more observing time without spending the additional "business" time required by most clubs. In fact, members of other astronomy clubs have joined the Tenn Valley Astronomy Club, while still maintaining their affiliation with their regular club, because it gives them additional observing opportunities and the chance to socialize with like-minded amateur astronomers.

Tri-State Astronomers

Washington County Planetarium
Commonwealth Avenue
Hagerstown, MD 21740
E-mail: solterra@intrepid.net
Members: 50

TRI-STATE ASTRONOMERS
Maryland • Pennsylvania • West Virginia

The Tri-State Astronomers hold monthly meetings at the Washington County Planetarium in Hagerstown, as well as monthly star parties during the new moon weekends, both at members' homes and at Greenbrier State Park, Boonsboro, Maryland. Public star parties are also given frequently and are at places like Antietam National Battlefield in Sharpsburg, Maryland. The club publishes *The Observer*, a monthly newsletter for members.

Dues: $15 annual (person/family)
       $10 annual (student)

Tulare Astronomical Association

9242 Avenue 184
Tulare, CA 93274
Mail:
P.O. Box 515
Tulare, CA 93274
Phone: (209) 686-0585
E-mail: kencope@lightspeed.net
Members: 120

The Tulare Astronomical Association has an observatory 8 miles south of Tulare, California. It is open to the public free of charge, and is staffed by volunteers.

Ufology Society International

P.O. Box 23103
RPO Connaught
Calgary, Alberta
Canada T2S-3B1
Phone: (403) 228-1804
E-mail: ufology@nucleus.com
Internet: www.ufology.org
Members: 500 plus

The Ufology Society International (USI) is an open and informal society for any ufology group, independent researcher, UFO enthusiast, scientist or skeptic who sees merit in seeking out and examining the available evidence, networking with others, and creating or collecting UFO related items of interest. USI is intended to encompass and promote the best the ufology community can offer. The Society offers an open and friendly forum for ufologists of all schools of thought, including scientists and skeptics.

USI offers an affordable and accessible Ufology Certificate Program and membership requires no monetary or service obligation. Basic lifetime membership is free to anyone who registers. Also, for a modest fee or by earning points, members are offered the opportunity to advance through four levels of self-instruction to become certified Field Representatives. Members who already possess adequate experience and ability may be approved for an existing position or one of their own creation. USI accepts and has members of all ages in numerous countries around the world. Special invitations are sent to selected individuals who have made a significant contribution in the field of Ufology. The Society is self-governed by members.

Victorian U.F.O. Research Society, Inc.

P.O. Box 43
Moorabbin, Victoria, Australia 3189
Phone: +61 3 9506 7080
E-mail: vufors@ozemail.com.au
Internet: www.ozemail.com.au/
~vufors
Members: 400

The Victorian U.F.O. Research Society was formed on February 17, 1957 and has held a dispassionate attitude on UFOs, claiming it is a scientific problem deserving closer attention. The Society meets regularly in general meetings and discussion nights and also publishes *The Australian UFO Bulletin*, a quarterly magazine.

Membership in this Society—which maintains the largest membership of any UFO organization in the Southern Hemisphere—is open to all who are genuinely interested in the subject.

Dues: $20 annual (US$, overseas membership)

Villanova Astronomical Society

Dept. of Astronomy & Astrophysics
Fourth Floor, Mendel Hall
800 Lancaster Ave
Villanova, PA 19085
Phone: (610) 519-7485
        or (610) 519-4820 after 5 p.m.
Fax: (610) 519-6132
E-mail: astronomy@ucis.vill.edu
**Internet: www.phy.vill.edu/astro/
vas/vas.htm**
Members: 35

*VAS operates the
Villanova Planetarium*

The Villanova Astronomical Society (VAS) consists of people drawn together by a common interest in the science of astronomy. The purpose of this society is to provide a medium through which all those interested in astronomy and its related fields may aid each other in developing their knowledge and ability in this science. Many of the members of the VAS are students in the undergraduate astronomy program at Villanova, but membership is open to the general community. The Society sponsor lectures, observing sessions, slides, movies, and field trips to enrich their understanding of the cutting edge topics in the field. Student lectures are generally held every other week on Wednesday evenings in Mendel Hall, room 350.

In the past two years, members of VAS have been involved in such projects as the detection and verification of an extra-solar planet orbiting CM Draconis, determining the chromo- spheric activity cycle of Proxima Centauri A&B, an empirical determination of the size of Mira A, analyzing the X-Ray background flux produced by RS CVn stars in the galactic halo, a first-ever accurate determination of the dust and gas abundance of carbon toward Tau CMa, and an extensive analysis of the physical and chemical properties of interstellar dust grains. Recently, VAS went on an observing run at the National Radio Astronomy Observatory in Green Bank, West Virginia. Data analysis and mapping of HII regions in the galactic plane continues.

VAS operates the Villanova Planetarium, which features a 6.1-meter dome and a Spitz A3P projector. Shows are available by appointment. The Society also operates the Villanova Observatory, which houses a 14-inch Schmidt-Cassegrain telescope for public observing and a 14-inch Schmidt-Cassegrain fitted with a ST-6 SBIG CCD with color filter wheel and spectrograph for use by members. The observatory is open during the school year from 8 p.m. until 10 p.m., Monday through Thursday (spring and winter hours are 7-9 p.m.) Call for special sessions and solar observing.

Dues: $5 per school year

Von Braun Astronomical Society

P.O. Box 1142
Huntsville, AL 35807
Phone: (205) 464-0945
E-mail: adamsml@pipeline.com
Internet: members.aol.com/
VBASTRSOC/VBAS.html
Members: 150

The Von Braun Astronomical Society (VBAS) was founded in 1954 as the Rocket City Astronomical Association (RCAA). During the formative years of the RCAA, guidance was provided by members of the von Braun missile team. Under their leadership, an observatory was completed in 1956. During the late fifties and early sixties, this observatory was used by Dr. von Braun and his family and even by Alan Shepard (the first U.S. astronaut in space) to plan for an eventual landing on the Moon. A planetarium was added in the 1960's and another observatory was completed in time for the return of Halley's comet in 1985. More recently, Dr. Ernst Stuhlinger, who was also a member of the von Braun team, constructed a solar telescope for sunspot observations. All facilities are located on a 13.5 acre tract of land in Monte Sano State Park (see the Von Braun Planetarium listing under Places of Interest for more information). In the early 1970's, the RCAA was renamed as the Von Braun Astronomical Society, to honor Dr. Wernher von Braun.

Warsaw Astronomical Society

1308 Sunset Ave.
Winona Lake, IN 46590
Members: 30
Internet: clubs.kconline.com/was/

Membership in the Warsaw Astronomical Society is open to anyone with an interest in astronomy. The club co-owns and operates the YMCA Camp Crosley/WAS Observatory near North Webster, Indiana. The observatory is open year round for qualified members, and for the public during special events and on clear Saturday evenings throughout the summer. The observatory is equipped for astrophotography, astrovideography, and CCD imaging.

Westchester Amateur Astronomers

P.O. Box 111
Chappaqua, NY 10514
Phone: (914) 734-1741
Internet: www.interport.net/
~justpat
Members: 120

Founded in 1983, the Westchester Amateur Astronomers meet the first Friday of the month under the dome of the Andrus Planetarium located at the Hudson River Museum, Yonkers, New York. A monthly public star party called *Starway to Heaven* is hosted at Ward Pound Ridge Reservation, Cross River, and a quarterly star party called *Moon Over Hastings* is held at Draper Park, Hastings, New York (Draper Park is the historic site of Henry Draper's Observatory Cottage).

Dues: $25 annual

Williamson County Astronomy Club

901 S. Church Street
Georgetown, TX 78626
Phone: (512) 838-1312
E-mail: wcacastro@aol.com
Members: 20

The Williamson County Astronomy Club meets the second Thursday of each month at 7:30 p.m. at the Bank-One in Round Rock, Texas. Activities include star parties, occasional special outings for meteor showers, comets, and other astronomical events, and lectures for area schools and organizations. The club is a member of the International Dark Sky Association. Membership benefits include discount rates on *Astronomy* magazine, *Sky & Telescope* magazine and other Sky Publishing Corporation products. There are no dues or fees.

Wragg Canyon Stargazers

5200 Wragg Canyon Road
Napa, CA 94558
Phone: (707) 966-0579
Members: 25

The Wragg Canyon Stargazers meet once per month for stargazing. The club has no fees or dues.

Organizations
QuickFind Index
A state by state listing

Alabama
Birmingham Astronomical Society
Mobile Astronomical Society
Von Braun Astronomical Society

Arizona
International Dark-Sky Association
Phoenix Astronomical Society

California
2111 Foundation for Exploration
Association of Lunar and Planetary
 Observers - Meteors Section
Astronomical Society of the Pacific
Astronomical Unit
Group 70, Inc.
International Association for
 Astronomical Art
International Meteor Organization
Kern Astronomical Society
Manteca Observing Group
Monterey Institute for Research
 in Astronomy
Mountain Skies Astronomical Society
Orange County Astronomers
Peninsula Astronomical Society
Planetary Society
Royal Cabrillo Astronomical
 Stargazing Society

Sacramento Valley Astronomical
 Society
San Diego Astronomy Association
San Jose Astronomical Association
San Mateo Astronomical Society
Sidewalk Astronomers
Stockton Astronomical Society
Tulare Astronomical Association
Wragg Canyon Stargazers

Colorado
Colorado Springs Astronomical
 Society
Northern Colorado Astronomical
 Society

Delaware
Delaware Astronomical Society

Florida
Ancient City Astronomy Club
Escambia Amateur Astronomers'
 Association
League of the New Worlds
Southeastern Association
 for Research in Astronomy

Georgia
Auburn Astronomical Society

Hawaii
Joint Astronomy Centre
Mauna Kea Astronomical Society

Idaho
Magic Valley Astronomical Society

Illinois
Astronomical League
Champaign-Urbana Astronomical
  Society and Observatory
Lake County Astronomical Society
Popular Astronomy Club, Inc.
Ryerson Astronomical Society
Skokie Valley Astronomers

Indiana
Evansville Astronomical Society
Kokomo Astronomy Club
Warsaw Astronomical Society

Iowa
Cedar Amateur Astronomers, Inc.
Central Iowa Astronomers

Kentucky
Louisville Astronomical Society, Inc.
National Deep Sky Observers Society
Northern Kentucky Astronomers

Louisiana
Pontchartrain Astronomy Society, Inc.
Shreveport-Bossier Astronomical
  Society, Inc.

Maine
Astronomical Society of Northern
  New England

Maryland
Greenbelt Astronomy Club
Harford County Astronomical Society
National Capital Astronomers
Patient Astronomers, Ltd.
Space Telescope Science Institute
Tri-State Astronomers

Massachusetts
Amherst Area Amateur
  Astronomers Association
Harvard-Smithsonian Center
  for Astrophysics
Springfield Stars Club
Students for the Exploration
  and Development of Space

Michigan
Astronomy Day Headquarters
Eastern Michigan University
  Astronomy Club
Friends of Astronomy On-Line
Kalamazoo Astronomical Society

Minnesota
NACOMM

Missouri
Astronomical Society of Kansas City

Montana
Southwest Montana Astronomical
  Society

Nebraska
Northeast Nebraska Astronomy Club
Omaha Astronomical Society

Nevada
Astronomical Society of Nevada

New Jersey
Amateur Astronomers, Inc.
Astronomical Society of the
  Toms River Area (ASTRA)
Hopatcong Area Amateur Astronomers
Morris Museum Astronomical Society
New Jersey Astronomical Association
North Jersey Astronomical Group

Rockland Astronomy Club of New York
SETI League, Inc.
Skyland Stargazers

New Mexico
Albuquerque Astronomical Society
Astronomical Society of Las Cruces
LodeStar Project
Pajarito Astronomers

New York
Amateur Observers' Society of
  New York, Inc.
Astronomical Society of Long Island
Space Frontier Foundation
Westchester Amateur Astronomers

North Carolina
Catawba Valley Astronomy Club
Center for the Study of
  Extraterrestrial Intelligence
Starfax Association

Ohio
Cincinnati Astronomical Society
Columbus Astronomical Society
Miami Valley Astronomical Society
Millstream Astronomy Club
Society of Amateur Radio Astronomers
Stillwater Stargazers

Oklahoma
Oklahoma City Astronomy Club

Oregon
Southern Oregon Skywatchers

Pennsylvania
Lackawanna Astronomical Society
Lehigh Valley Amateur
  Astronomical Society
Villanova Astronomical Society

Rhode Island
Skyscrapers, Inc.

South Carolina
Midlands Astronomy Club
North American Meteor Network

Tennessee
Tenn Valley Astronomy Club

Texas
Abilene Astronomical Society
Amarillo Astronomy Club
El Paso Astronomy Club
Mutual UFO Network, Inc.
Williamson County Astronomy Club

Virginia
Back Bay Amateur Astronomers
Northern Virginia Astronomy Club
Skywatchers Astronomy Club
Society for Scientific Exploration

Washington
Amateur Telescope Makers
  Association
Battle Point Astronomical Association
Olympic Astronomical Society
Seattle Astronomical Society

Washington, D.C.
American Astronomical Society
Association of Universities for
  Research in Astronomy
NASA Headquarters
National Space Society

West Virginia
Kanawha Valley Astronomical Society
Ohio Valley Astronomical Society

Wisconsin
Madison Astronomical Society
Neville Public Museum
  Astronomical Society
North East Wisconsin Stargazers
  (NEWSTAR)
Sheboygan Astronomical Society

Wyoming
Campbell County High School
  Astronomy Club
Central Wyoming Astronomical
  Society
Cheyenne Astronomical Society
Jackson (Hole) Astronomy Club

Australia
Astronomical Society of Australia
Victorian U.F.O. Research Society, Inc.

Canada
Comox Valley Astronomy Club
Ufology Society International

France
European Space Agency

Internet
Galaxies Astronomy Club
Friends of Astronomy On-Line

United Kingdom
British National Space Centre (BNSC)
Royal Greenwich Observatory

Appendix A
Astronomical League Societies

*The following organizations are members
of the Astronomical League*
(ordered by state)

Alabama

Auburn Astronomical Society
Mr. John B. Zachry
P. O. Box 778
West Point, GA 31833-0778
Phone: 706-645-1680
E-mail:
    rwhigham@mont.mindspring.com
Internet: www.mindspring.com/
    ~rwhigham/aas_home.htm

Birmingham Astronomical Society
P. O. Box 36311
Birmingham, AL 35236
Internet: www.secis.com/
    ~braeunig/bas.html
Contact: Laura C. Phelps
1494 Milner's Crescent, #G
Birmingham, AL 35205
Phone: 205-933-6729

Mobile Astronomical Society
6101 Girby Road
Mobile, AL 36693
Contact: Loxley Greaves
5659 Diane Drive
Mobile, AL 36618
Phone: 334-343-0255

Von Braun Astronomical Society
P. O. Box 1142
Huntsville, AL 35807
Internet: members.aol.com/
    VBASTROSOC/VBAS.html
Contact: Michael Cowger
2007 Greenwood Place
Huntsville, AL 35802
Phone: 205-880-7241

Alaska

Astronomical Units
Contact: Robert Fischer
Box 82210
Fairbanks, AK 99708
Phone: 907-456-6586

Arizona

Phoenix Astronomical Society
Contact: Terri Renner
2750 E. Larkspur Drive
Phoenix, AZ 85032
Phone: 602-971-3355
E-mail: starstuf@aztec.asu.edu
Internet: www.netzone.com/~ranger/
    pas/pastime.htm

**Tucson Amateur
Astronomy Association**
P. O. Box 41254
Tucson, AZ 85717
Internet: www.primenet.com/
　　　　~lwilson/taaa/taaa.html
Contact: Bob Gent
3661 N. Round Rock Drive
Tucson, AZ 85750
Phone: 520-721-5060
E-mail: RLGent@aol.com

Arkansas

**Astronomical Society of
NW Arkansas**
Contact: Eliot Neel
2300 Old Wire Road
Fayetteville, AR 72703
Phone: 501-443-9127

**Central Arkansas
Astronomical Society**
Contact: John Reed
10 Kirkland Dr.
Cabot, AR 72023
Phone: 843-8796
Internet: www.business.uca.edu/
　　　partners/caas/caashome.html

Red River Astronomy Club
305 Anglewood Drive
Nashville, AR 71852
Contact: Pat Rogers
211 County Line Road N.
Nashville, AR 71852
Phone: 501-845-4683

California

Central Valley Astronomers
P.O. Box 3063
Pinedale, CA 93650
Phone: 209-435-8232
E-mail: CVAfresno@aol.com
Internet: members.aol.com/cvafresno/
　　　index.html
Contact: David Lehman
40 E. Minarets, #10
Fresno, CA 93650
E-mail: DLehman111@aol.com

Eastbay Astronomical Society
4917 Mountain Blvd.
Oakland, CA 94619
E-mail: eas@cosc.org
Internet: www.cosc.org/EAS.html
Contact: Carter W. Roberts
5 Highgate Court
Berkeley, CA 94707
Phone: 510-524-2146
E-mail: croberts@cosc.org

Lake Tahoe Astronomers
P. O. Box 14271
S. Lake Tahoe, CA 96151-4271
Contact: Tim Torpin
P. O. Box 612632
S. Lake Tahoe, CA 96152-2632
Phone: 916-541-6949

Mt. Diablo Astronomical Society
2466 Sky Road
Walnut Creek, CA 94596
Contact: Joel Goodman
2 Miramonte Drive
Moraga, CA 94556
Phone: 510-631-0829

San Bernardino Valley Amat.uer Astronomers

c/o G. F. Beattie Planetarium
SBVC, 701 S. Mount Vernon Avenue
San Bernardino, CA 92410
Contact: Chris Clarke
249 E. Van Koevering
Rialto, CA 92376
Phone: 909-875-6694

San Jose Astronomical Association

James Van Nuland
3509 Calico Avenue
San Jose, CA 95124
Phone: 408-371-1307
Internet: www.rahul.net/resource/sjaa

South Bay Astronomical Society

c/o Microcosm, Inc.
2733 Crenshaw Blvd., Suite 350
Torrance, CA 90501
E-mail: jwertz@smad.com
Internet: www.meteorite.com/sbas/
Contact: Dave Wright
5501 Norton Street
Torrance, CA 90503
Phone: 310-542-6207
E-mail: DaveW70000@aol.com

Temecula Valley Astronomers

Contact: Selma Lesser
38212 Calle Jojoba
Temecula, CA 92592
Phone: 909-699-9637

Ventura County Astronomical Society

P. O. Box 982
Simi Valley, CA 93062
E-mail: vcas1@aol.com
Internet: www.serve.net/vcas/vcas.html
Contact: Tim Robertson
Phone: 805-584-6706
E-mail: trobert@earthlink.net

Western Observatorium

4141 Ball Road, Suite 212
Cypress, CA 90630-3400
Internet: www.cogent.net/~pyrrho/
    wohome.htm
Contact: Paul M. Livio
2889 West 230th Street
Torrance, CA 90505-2858
Phone: 310-534-8132

Colorado

Colorado Springs Astronomical Society

Box 62022
Colorado Springs, CO 80962
E-mail: mfrazier@us.oracle.com
Contact: Joy Simington
265 Winding Meadow Way
Monument, CO 80132
Phone: 719-481-8169
E-mail: XARA@kktv.com

Denver Astronomical Society

Chamberlin Observatory
2930 E. Warren Avenue
Denver, CO 80208
Internet: www.du.edu/~pryan/das.html.
Contact: Debra L Davis
8749 W Cornell Ave #6
Lakewood, CO 80227-4813
Phone: 303-988-3586
E-mail: saturna@ix.netcom.com

Longmont Astronomical Society

Contact: Bob Spohn
1342 Garden Place
Longmont, CO 80501
Phone: 303-772-1470
E-mail: 72274.415@compuserve.com
Internet: www.fsl.noaa.gov/frd-bin/
    albers_las.homepage.cgi

Western Colorado Astronomy Club
Aaron Reid
816 Jamaica
Grand Junction, CO 81506
Phone: 970-245-4995

Connecticut

Mattatuck Astronomical Society
Contact: Chet Case, Jr.
17 Abbott Avenue
Terryville, CT 06786
Phone: 203-589-5625
E-mail: btbrick@mail.snet.net

Westport Astronomical Society
E-mail: jsoisson@world.std.com
Internet: www.was.visionnet.com
Contact: John Soisson
86 Arthur Dr
South Windsor, CT 06074-3646
Phone: 203-648-0837
E-mail: jsoisson@world.com

District of Columbia

Northern Virginia Astronomy Club
Internet: astro.gmu.edu/~novac
Contact: Bill Jensen
7405 Ridge Oak Ct.
Springfield, VA 22153
Phone: 703-866-1380
E-mail: william.jensen@tcs.wap.org

Florida

Ancient City Astronomy Club
P. O. Box 546
St. Augustine, FL 32085-0546
Contact: A. L. Ponjee
P. O. Box 354030
Palm Coast, FL 32135-4030
Phone: 904-446-4338

Brevard Astronomical Society
Contact: Chuck K. Adams
1413 Crest Avenue
Titusville, FL 32780
Phone: 407-452-4395

Escambia Amateur Astronomers' Association
Internet: www.pen.net/~mew/eaaa/index.htm
Contact: Wayne Wooten
6235 Omie Circle
Pensacola, FL 32504
Phone (voice mail): 904-484-1152

Everglades Astronomical Society
Contact: Mr. Bernard Wahl
754 101st Avenue N.
Naples, FL 33963
Phone: 941-597-3645
E-mail: bernard163@aol.com

Local Group of Deep Sky Observers
Contact: Vic Menard
2311 23rd Avenue West
Bradenton, FL 34205
Phone: 813-747-8334

Northeast Florida Astronomical Society
Internet: www.jaxnet.com/~rcurry/nefas.html
Contact: Roger D. Curry
2321 Camden Avenue
Jacksonville, FL 32207
Phone: 904-398-1335
E-mail: rcurry@jaxnet.com

**Southern Cross
Astronomical Society**
Internet: www.mangonet.com/scas
Contact: Ms. Barbara W. Yager
690 Allendale Road
Key Biscayne, FL 33149
Phone: 305-361-2502

Tallahassee Astronomical Society
Internet: www.polaris.net/~tas/tas.html
Contact: Conrad Scott Howard
7545 Old St. Augustine Road
Tallahassee, FL 32311
Phone: 904-878-6236
E-mail: cshoward@freenet.tlh.fl.us

**Treasure Coast
Astronomical Society**
Contact: Charlotte A. Bilder
1755 NW Fork Road
Stuart, FL 34994
Phone: 561-692-1124

Georgia

Astronomy Club of Augusta
P. O. Box 6373
Augusta, GA 30916-6373
Contact: Jerry Barton
3601 Spanish Court
Hephzibah, GA 30815
Phone: 706-793-3420

Athens Astronomical Society
Contact: Maurice E. Snook
160 Plantation Drive
Athens, GA 30605
Phone: 706-543-3753

Atlanta Astronomy Club
Suite A9-305
3595 Canton Road
Marietta, GA 30066
Internet: stlspb.gtri.gatech.edu/astrotxt/
atlastro.html
Contact: Leonard Abbey
1002 Citadel Drive
Atlanta, GA 30324
Phone: 404-634-1222
E-mail: labbey@mindspring.com

**Middle Georgia
Astronomical Society**
Contact: Mike Hood
110 Van Drive
Kathleen, GA 31047
Phone: 912-987-0175

Idaho

Boise Astronomical Society
Contact: Bret Blakeslee
722 Pearl Street
Boise, ID 83705
Phone: 208-385-7636

Idaho Falls Astronomical Society
Contact: James F. Ruggiero
1710 Claremont Lane
Idaho Falls, ID 83404
Phone: 208-524-6317
E-mail: ruggi@srv.net

Magic Valley Astronomical Society
Contact: Forrest Ray
531 Filer Avenue, Apt. 3A
Twin Falls, ID 83301
Phone: 208-736-8678

Panhandle Astronomers, Ltd.
P. O. Box 2051
Hayden, ID 83835
Contact: Jim Mullen
23 E. Falcon Avenue
Spokane, WA 99218
Phone: 509-467-8073

Illinois

Chicago Astronomical Society
P. O. Box 30287
Chicago, IL 60630-0287
Phone: 312-725-5618
Contact: Mr. Michael F. Barrett
5735 W. Pensacola Ave.
Chicago, IL 60634-1722
Phone: 773-725-3961

Fox Valley Astronomical Society
P. O. Box 508
Batavia, IL 60510
Contact: David Johnson
1174 Pine Street
Batavia, IL 60510
Phone: 630-406-6128

Fox Valley Skywatchers
Contact: Russ Maxwell
164 Hilltop Lane
Sleepy Hollow, IL 60118
Phone: 847-428-4178
E-mail: maxr7@aol.com

Naperville Astronomical Association
Contact: Kenneth Timmons
69 Sheffield
Montgomery, IL 60538
Phone: 630-896-2289

Northern Illinois Stargazers
Contact: Kerry Tumleson
217 S. Dement Avenue
Dixon, IL 61021
Phone: 815-288-5940

Northwest Suburban Astronomers
E-mail:
102127.2666@compuserve.com
Internet: ourworld.compuserve.com/
homepages/JohnP_Dwyer
Contact: R. K. Radhakrishnan
741 N. Walnut Lane
Schaumburg, IL 60194
Phone: 847-885-4874

Peoria Astronomical Society
E-mail: dware@bradley.bradley.edu
Internet: bradley.bradley.edu/~dware/
index.html
Contact: Scott Swords
7411 N. Patton Lane
Peoria, IL 61614
E-mail: sswords@iaonline.com

Popular Astronomy Club
Contact: Lee Farrar
2232 24th Street
Rock Island, IL 61201
Phone: 309-786-6844
E-mail: lmfastro@aol.com

Quad-Cities Astronomical Society
P. O. Box 3706
Davenport, IA 52808
E-mail: saber@revealed.net
Contact: Steve J. Martens
5021 46th Ave. Ct.
Moline, IL 61265
Phone: 309-797-2712

Rockford Amateur Astronomers, Inc.
Contact: Barry Beaman
6804 Alvina Road
Rockford, IL 61101
Phone: 815-962-6540
E-mail: Praesepe44@aol.com

Skokie Valley Astronomers
Internet: ourworld.compuserve.com/
    homepages/e_neuzil
Contact: Gretchen Patti
1 South 751 Avon
Warrenville, IL 60555
Phone: 630-393-7929
E-mail: gretchen@adichi.com

Twin City Amateur Astronomers
E-mail: mprogers@math.ilstu.edu
Contact: Sandra McNamara
P.O. Box 258
Stanford, IL 61774
Phone: 309-379-2751
E-mail: SandyMc456@aol.com

Indiana

Evansville Astronomical Society
Contact: Scott Conner
10119 Fischer Road
Evansville, IN 47720
Phone: 812-963-9110
E-mail: 76573.1213@compuserve.com

Fort Wayne Astronomical Society
P. O. Box 11093
Fort Wayne, IN 46855
Contact: Larry Clifford
428 Wallen Hills Drive, #2
Fort Wayne, IN 46825
E-mail: 75544.3011@compuserve.com

Michiana Astronomical Society
Internet: michiana.org/MFNetSci/MAS/
    index.html
Contact: Dan Smith
2904 Hilltop Drive
South Bend, IN 46614
Phone: 219-282-1086

Wabash Valley Astronomical Society
E-mail: wyncott@purdue.edu
Internet: www.holli.com/~wbormann/
    wvas.htm
Contact: Mike Marsh
7509 North 75 East
West Lafayette, IN 47906
Phone: 317-497-0056
E-mail: mhm@cc.purdue.edu

Warsaw Astronomical Society
E-mail: michael.beaver
    @virtualofficesystems.com
Internet: clubs.kconline.com/was
Contact: James D. Tague
1308 Sunset Drive
Winona Lake, IN 46590
Phone: 219-269-1856

Iowa

Ames Area Amateur Astronomers
3912 Brookdale Circle
Ames, IA 50010
Internet: www.cnde.iastate.edu/
    aaaa.html
Contact: Jim Bonser & Family
1009 Harding Street
Tama, IA 52339
Phone: 515-484-5235
E-mail: jbonser@aol.com

Cedar Amateur Astronomers, Inc.
E-mail: dslauson@cedar-rapids.net
Internet: www.geocities.com/
    CapeCanaveral/8866/
    CAA-home.html
Contact: Keith Sippy
4586 Sutton Road
Central City, IA 52214
Phone: 319-438-6033
E-mail: keith.sippy@cedar-rapids.net

Central Iowa Astronomers
2305 Drake Park Avenue
Des Moines, IA 50311-4312
Contact: Joanne Hailey
1116 42nd Street
Des Moines, IA 50311
Phone: 515-277-2739

Des Moines Astronomical Society
Contact: Miss Connie Allen
2307 49th Street
Des Moines, IA 50310-2538

Quad-Cities Astronomical Society
P. O. Box 3706
Davenport, IA 52808
E-mail: saber@revealed.net
Contact: Steve J. Martens
5021 46th Ave. Ct.
Moline, IL 61265
Phone: 309-797-2712

Southeastern Iowa Astronomy Club
E-mail: seiac@aol.com
Contact: Evan Bachtell
109 Spring
W. Burlington, IA 52655
Phone: 319-752-6030
E-mail: ebachtell@aol.com

Kansas

Kansas Astronomical Observers
E-mail: pgoedken@wichita.fn.net
Internet: www.feist.com/~pgoedken/
    kao.html
Contact: Chuck Glascock
412 Garst
Wichita, KS 67209
Phone: 316-945-8654

**Kansas Astrophotographers &
Observers Society**
Contact: Rob Robinson
515 W. Kump
Bonner Springs, KS 66012
Phone: 913-422-1262
E-mail: robinson@sky.net

Salina Astronomy Club
Contact: Keith Rawlings
803 Mike Drive
Salina, KS 67401
Phone: 913-827-6004

Kentucky

Blue Grass Astronomical Society
Contact: Dr. Bradley C. Canon
1016 Della Drive
Lexington, KY 40504
Phone: 606-278-6155
E-mail: pol140@ukcc.uky.edu

Cincinnati Astronomical Society
Contact: Jonathan Jennings
10096 Golden Pond Dr.
Union, KY 41091
Phone: 606-384-7108
E-mail: jonj@usa.pipeline.com

Louisville Astronomical Society
P. O. Box 20742
Louisville, KY 40250-0742
Internet: www.venus.net/~dhaggard/
Contact: Chuck Allen
1007 Rollingwood Lane
Goshen, KY 40026
Phone: 502-228-3043
E-mail: 74023.2331@compuserve.com

West Kentucky
Amateur Astronomers
Contact: Sharon Steedly
955 Barnhill Road
Providence, KY 42450
Phone: 502-667-7195

Louisiana

Baton Rouge Astronomical Society
E-mail: fred@mail.advtel.net
Internet: www.eatel.net/~fred/bras/
    bras.html
Contact: Craig Brenden
6348 Double Tree Court
Baton Rouge, LA 70817-8915
Phone: 504-751-1685

Pontchartrain Astronomy Society
Contact: Bob Sylvester
13 Idlewood Place
New Orleans, LA 70123-1525
Phone: 504-738-2934
E-mail: whbk80a@prodigy.com

Shreveport-Bossier Astronomical
Society, Inc.
353 Ockley Drive
Shreveport, LA 71105
Internet: www.lsus.edu/nonprofit/sbas/
Contact: Terry Atwood
3448 Johnette
Shreveport, LA 71105
Phone: 318-868-3581

Maine

45th Parallel Amat. Astronomers of
Maine
RD #1, Box 1170
Exeter, ME 04435
E-mail: kthhoward@aol.com
Contact: William Lashon
P.O. Box 142
Skowhegan, ME 04976-0142
Phone: 207-474-7200
E-mail: tabor@tabor.sdi.agate.net

Astronomical Society of
Northern New England
East Sky Observatory
P. O. Box 497A
Kennebunkport, ME 04046
Phone: 207-967-5945
E-mail: astronomy@nlis.net
Internet: web.nlis.net/~mesky/
Contact: Peter Talmage
P. O. Box 497A
Kennebunkport, ME 04046-0497
Phone: 207-967-5945
E-mail: talmageeng@cyberTours.com

Galileo Society
Contact: David Gay
RR#2, Box 1200
Livermore Falls, ME 04254

Maryland

Baltimore Astronomical Society
Contact: Jim Nickel
502 E. 41st Street
Baltimore, MD 21218-1210
Phone: 410-323-0439

Cumberland Astronomy Club
Contact: Dr. Stephen Luzader
59 Centennial Street
Frostburg, MD 21532
Phone: 301-689-1976
E-mail: e2ppluz@fre.fsu.umd.edu

Goddard Astronomy Club
E-mail: jeff.guerber@gsfc.nasa.gov
Contact: Robert Dutilly
8676 Mission Road
Jessup, MD 20794
Phone: 301-286-4916;
E-mail: bob.dutilly@gsfc.nasa.gov

Greenbelt Astronomy Club
Contact: Douglas L. Love
3-D Plateau Place
Greenbelt, MD 20770

Harford County Astronomical Society
Contact: Joyce Bish
1205 Wild Orchid Drive
Fallston, MD 21047
Phone: 410-877-1986

Northern Virginia Astronomy Club
Internet: astro.gmu.edu/~novac
Contact: Bill Jensen
7405 Ridge Oak Ct.
Springfield, VA 22153
Phone: 703-866-1380
E-mail: william.jensen@tcs.wap.org

Southern Maryland Astronomical Society
P. O. Box 1727
White Plains, MD 20695-1727
Contact: George Bustilloz
Phone: 301-743-9161
E-mail: bustillo@ix.netcom.com

Tri-State Astronomers
E-mail: jimstan@bbs.kis.net
Contact: Robert Johnsson
7422 Ridge Road
Frederick, MD 21702
Phone: 301-371-5215

Westminster Astronomical Society
Contact: Frank Filemyr
2112 Woodview Road
Finksburg, MD 21048
Phone: 410-876-1924

Massachusetts

Aldrich Astronomical Society
Contact: Ed Guries
14 Wyoma Drive
Auburn, MA 01501
Phone: 508-799-0184

Amherst Area Amateur Astronomers Association
Contact: John N. Davis
177 N. Pleasant Street, #11
Amherst, MA 01002-1720
Phone: 413-256-0578

North Shore Amateur Astronomy Club
c/o David Thomas
12 Claire Road
Amesbury, MA 01913
Internet: www.star.net/people/~ntreal/
nsaac.htm
Contact: Paul Graveline
9 Stirling Street
Andover, MA 01810
Phone: 508-470-1971

Springfield Stars Club
Internet: ourworld.compuserve.com/
     homepages/efaits
Contact: Richard Sanderson
P. O. Box 2793
Springfield, MA 01101
Phone: 413-267-5283

Michigan

Amateur Astronomers of Jackson
1016 Westfield Drive
Jackson, MI 49203-3630
Contact: John Isles
1016 Westfield Drive
Jackson, MI 49203
Phone: 517-789-8697
E-mail: jisles@voyager.net

Capital Area Astronomy Club
E-mail: dbatch@msu.edu
Internet: www.pa.msu.edu/abrams/
     caac.html
Contact: Brian Cerveny
816 Summit Street
Owosso, MI 48867

Delta Astronomical Society
Contact: Bob Jasmund
907 4th Avenue So.
Escanaba, MI 49829
Phone: 906-786-8114

Ford Amateur Astronomy Club
P. O. Box 7527
Dearborn, MI 48121-7527
Contact: Brian Gossiaux
6824 Foxthorn
Canton, MI 48187
Phone: 313-274-3416

**Grand Rapids Amateur
Astronomers Association**
3308 Kissing Rock Road SE
Lowell, MI 49331
E-mail: graaa@grfn.org
Internet: www.grfn.org/~graaa
Contact: Dan Braybrook
3756 E. Omaha
Grandville, MI 49418
Phone: 616-531-4971
E-mail: dlb525@aol.com

Kalamazoo Astronomical Society
c/o KAMSC
600 West Vine, Suite 400
Kalamazoo, MI 49008
E-mail: MJDUPUIS@am.pnu.com
Internet: www.geocities.com/
     CapeCanaveral/Lab/1000/kas.html
Contact: Mr. Mike Dupuis
46041 CR 652
Mattawan, MI 49071
Phone: 616-668-4930

Marquette Astronomical Society
Shiras Planetarium
1201 W. Fair Avenue
Marquette, MI 49855
Internet: www.mich.com/~bhalbroo/
     bhalbroo.html.
Contact: Stephen H. Peters
908 W. Magnetic St.
Marquette, MI 49855
Phone: 906-228-6636
E-mail: speters@nmu.edu

**McMath-Hulbert
Astronomical Society**
895 N. Lake Angelus Road
Lake Angelus, MI 48326
Contact: Marty Kunz
29836 Hillbrook
Livonia, MI 48152
Phone: 810-477-0546
E-mail: aj130@detroit.freenet.org

Oakland Astronomy Club
Contact: Chuck Bovee
1370 Edgeorge
Waterford, MI 48327
Phone: 810-673-9638

**Shoreline Amateur
Astronomy Association**
Internet: www.macatawa.org/~saaa/
Contact: Mr. Robert Wade
3882 62nd Street
Holland, MI 49423
Phone: 616-396-3614
E-mail: wader@macatawa.org

Warren Astronomical Society
P. O. Box 1505
Warren, MI 48090-1505
Internet: www.eaglequest.com/
~bondono/WAS/iwas.html
Contact: Ben Tolbert
20206 Vermander
Clinton Twp., MI 48035
Phone: 810-790-8292

Minnesota

Fargo-Moorhead Astronomy Club
MSU Planetarium
Moorhead State University
Moorhead, MN 56563
Contact: Frank H. Clark
1608 4th Street N.
Fargo, ND 58102-2304
Phone: 701-280-9689
E-mail: 74244.2533@compuserve.com

Minnesota Astronomical Society
c/o Science Museum of Minnesota
30 E. 10th Street
St. Paul, MN 55101
Internet: www.geom.umn.edu/
~slevy/mas/
Contact: Stuart Levy
3540 33rd Ave. S.
Minneapolis, MN 55406-2725
Phone: 612-729-9289
E-mail: slevy@geom.umn.edu

Willmar Area Astronomy Club
622 9th Street SW
Willmar, MN 56201
Contact: Paul Nelson
504 Sunrise Drive
Olivia, MN 56277
Phone: 612-523-1781

Mississippi

Jackson Astronomical Association
P. O. Box 766
Clinton, MS 34960
Internet: www2.netdoor.com/~hboswell/
jaa/jaa.html
Contact: Larry Luke
210 Creekline Drive
Madison, MS 39110
Phone: 601-856-5979
E-mail: 76665.3160@compuserve.com

Rainwater Astronomical Association
Contact: Wayne Coskrey
P. O. Box 549
Starkville, MS 39759-0549
Phone: 601-323-4944

Missouri

Astronomical Society of Kansas City
P. O. Box 400
Blue Springs, MO 64015
Internet: www.sound.net/~askc/
Contact: Betty Iorg
7241 Jarboe
Kansas City, MO 64114
Phone: 444-4878
E-mail: ciorg@sound.net

Central Missouri Amateur Astronomy Club
Contact: Paul Rothove
16065 S. Palis Nichols Road
Hartsburg, MO 65039
Phone: 314-657-1314

Eastern Missouri Dark Sky Observers
Contact: Judy Kemp
8247 Hwy YY
New Haven, MO 63068
Phone: 573-459-6633

Lincoln County Stargazers
Jim Marsh, Secretary
341 Monroe Street
Troy, MO 63379
Contact: Jim & Altha Marsh
341 Monroe Street
Troy, MO 63379
Phone: 314-528-6702

McDonnell-Douglas Amateur Astronomy Club
Contact: Francis Baum
7320 Chamberlain Ave
University City, MO 63130
Phone: 314-726-2375
E-mail: baum@mpsn01.mdc.com

Pythagorica Astronomical Society of Jefferson County
Jefferson Memorial Hospital
P. O. Box 350
Crystal City, MO 63019-0350
Phone: 314-933-1000 ext. 1620
Contact: Rosie & Chris Klossner
7050 Three B Road
Cedar Hill, MO 63016
Phone: 314-274-4628

St. Louis Astronomical Society
E-mail: scopes777@aol.com
Contact: Bruce Logan
8125 S. Laclede Station Road
St. Louis, MO 63123-2114
Phone: 314-843-2595

Nebraska

Northeast Nebraska Astronomy Club
710 So. 4th
Norfolk, NE 68701
Contact: Gregg Adams
710 So. 4th
Norfolk, NE 68701
Phone: 402-371-0891

Omaha Astronomical Society
Internet: www.top.net/cdcheney/
index.html
Contact: Rich Merten
2918 Bridgeford Road
Omaha, NE 68124
Phone: 402-390-9053
E-mail: rmerten@aol.com

Prairie Astronomy Club of Lincoln
Internet: www.infoanalytic.com/pac/
index.html
Contact: Dr. L. Lee Thomas
5827 LaSalle St.
Lincoln, NE 68516
Phone: 402-483-5639

New Jersey

Amateur Astronomers, Inc.
Contact: Mr. George Chaplenko
73 Alexander Street
Edison, NJ 08820
Phone: 908-549-0615

Astronomical Society of Toms River Area
Contact: R. Erik Zimmermann
Planetarium - Ocean County College
Toms River, NJ 08754-2001
Phone: 908-255-0343
E-mail: zimmermann@monmouth.com

Rockland Astronomy Club
73 Haring Street
Closter, NJ 07824-1709
E-mail: durban@aol.com
Contact: C. Thomas Massey
14 Iona Place
Glen Rock, NJ 07452
Phone: 201-447-2581

S.T.A.R. Astronomy Society
Internet: www.monmouth.com/~ksears
Contact: Donald Odegard
13 Carolina Avenue
Port Monmouth, NJ 07758
Phone: 908-787-4766

Sheep Hill Astronomical Association
Contact: Eric Jacoves
13 Seneca Avenue
Rockaway, NJ 07866
Phone: 201-625-1768

Springfield Telescope Makers
Internet: www.ilinkgn.net/commercl/author/stelfane.htm
Contact: Margaret Salter
1135 Applewood Drive
Freehold, NJ 07728-3979
Phone: 908-462-0875

New Mexico

Alamagordo Amateur Astronomers
Contact: Paul Carnes
1502 Jefferson Avenue
Alamagordo, NM 88310
Phone: 505-437-4505
E-mail: paul.carnes@mdcbbs.com

Clovis Astronomy Club
Contact: Carol Nash
3517 Corlington Lane
Clovis, NM 88101
Phone: 505-763-7455

Southeast New Mexico Astronomy Club
Robert H. Goddard Planetarium
11th & North Main
Roswell, NM 88201
Contact: Karen B. Nelson
610 Mission Arch Drive
Roswell, NM 88201
Phone: 505-623-5781

New York

Albany Area Amateur Astronomers
Contact: Raymond Bogucki
253 Hill Hollow Road
Petersburg, NY 12138
Phone: 518-658-3138
E-mail: rfbogucki@aol.com

Amateur Observers Society of New York

E-mail: charlie087@aol.com
Internet: members.aol.com/
 rjbenjamin/aos.html
Contact: Bob Godfrey
67 Fulton Street
Massapequa Park, NY 11762
Phone: 516-799-4885

Astronomy Section Rochester Academy of Science

E-mail: marty.xkeys@xerox.com
Contact: Kenneth & Trudie Brown
1471 Long Pond Road, Apt. 305
Rochester, NY 14626-4131
Phone: 716-889-4299

Mohawk Valley Astronomical Society

E-mail: rsomer@hamilton.edu
Contact: Arlene & Richard Somer
7 Miller Road
Clinton, NY 13323
Phone: 315-853-6056

North Carolina

Cape Fear Astronomical Society

Contact: Doug Greene
5807 Dorothy Avenue
Wilmington, NC 28403
Phone: 910-799-2255
E-mail: 70400.1331@compuserve.com

Cleveland County Astronomical Society

Williams Observatory
Gardner-Webb University
Boiling Springs, NC 28017
Contact: Tom English
Phone: 704-434-4433

Greensboro Astronomy Club

E-mail: dennish522@aol.com
Contact: Chris King
6907 Ridge Haven Drive
Greensboro, NC 27410
Phone: 910-668-4502

Raleigh Astronomy Club, Inc.

P. O. Box 10643
Raleigh, NC 27605
E-mail: rac@rtpnet.org
Internet: RTPnet.org/~rac/
Contact: Phyllis Lang
811 Roanoke Drive
Cary, NC 27513

Thomasville Area Astronomy Club

E-mail: sfzl65a@prodigy.com
Contact: Bud Oates
305 Gregg St
High Point, NC 27263
Phone: 910-431-5062
E-mail: oatesla@hpe.infi.net

Ohio

Miami Valley Astronomical Society

Internet: www.mvas.org
Contact: Ms. Keneil Blaho
8230 W. Rt. 718
Pleasant Hill, OH 45359-9701
Phone: 937-676-5891
E-mail: blaho@erinet.com

Mid-Western Astronomers

Contact: Susan Rismiller
5760 Richland Circle
Milford, OH 45150
Phone: 513-248-1577
E-mail: steve.rismiller@prodigy.com

Millstream Astronomy Club
Contact: Roger L. Myers
568 Scott Street
Lima, OH 45804
Phone: 419-221-2248
E-mail: rogergop@brutus.bright.net

**Northern Ohio Valley
Astronomy Educators**
Contact: Ralph J. Hadley, Jr.
57815 Spring Hill Road
Bellaire, OH 43906
Phone: 614-676-5659

Northwest Ohio Visual Astronomers
Contact: Frank Myers
25570 Brittany Road
Perrysburg, OH 43551
Phone: 419-872-0910;
E-mail: telescope7@aol.com

Oil Region Astronomical Society, Inc.
Contact: Steve & Debbie Behringer
256 Buckeye Circle
Columbus, OH 43217
Phone: 614-492-1521
E-mail:
    102051.3276@compuserve.com

Richland Astronomical Society
Internet: www.mfdonline.com/
    wro/wro.html
Contact: Keith A. Moore
331 S. Market Street
Galion, OH 44833
Phone: 419-468-3542
E-mail: kamoore@richnet.net

**Sandusky Valley Amateur
Astonomy Club**
Contact: Thomas Fretz
650 S. Washington St.
Tiffin, OH 44883
Phone: 419-448-9377

Stillwater Stargazers
Bruckner Nature Center
5995 Horseshoe Bend Road
Troy, OH 45373
E-mail: garypike@bright.net
Internet: www.qns.com/~berrys/
    astro/astro.htm
Contact: Carl McClure
7041 W. Cox Road
Pleasant Hill, OH 45359
Phone: 513-676-2879

Oklahoma

Astronomy Club of Tulsa
E-mail: dem@galstar.com
Contact: Newell Pottorf
3832 S. Victor Avenue
Tulsa, OK 74105
Phone: 918-742-7577

Bartlesville Astronomical Society
Contact: Ken Willcox
Rt. 2, Box 940
Bartlesville, OK 74006
Phone: 918-333-1966

**Northwest Oklahoma
Astronomy Club**
Contact: Dan Mathews
1719 Pawnee
Enid, OK 73703
Phone: 405-233-5707

Odyssey Astronomy Club
Steve Atherton
5606B S.E. 15th
Midwest City, OK 73110
Contact: Steve Arthurton
5606B S.E. 15th
Midwest City, OK 73110
Phone: 405-732-1350

Oklahoma City Astronomy Club
Contact: Beryl Cadle
P. O. Box 283
Bethany, OK 73008
Phone: 405-789-3742
E-mail: ccadle@juno.com

TUVA Astronomy Organization
Contact: Ron Wood
Rt. 3, Box 638
Checotah, OK 74426
Phone: 918-474-3275

Oregon

Eugene Astronomical Society
E-mail: hartw@efn.org
Internet: www.efn.org/~bsackett
Contact: Larry Dunn
34625 Seavey Loop Road
Eugene, OR 97405
Phone: 541-747-9719

Rose City Astronomers
Internet: www.teleport.com/~rca/
Contact: Dale W. Fenske
16139 NE Siskiyou
Portland, OR 97230
Phone: 256-1840
E-mail: fenske@uofport.edu

Southern Oregon Skywatchers
880 Glendower Street
Ashland, OR 97520
E-mail: galaxygl@mind.net
Contact: Joe Porhammer
1090 Tunnel Road
Glendale, OR 97442
Phone: 541-832-2614

Pennsylvania

Beaver Valley Astronomy Club
Contact: Paul Vondra
36 S. Harrison Avenue, #27
Bellevue, PA 15202
Phone: 412-734-0653

Berks County Amateur Astronomers
P. O. Box 6150
Wyomissing, PA 19610
E-mail:
    104663.1600@compuserve.com
Internet: www.voicenet.com/~bcaas
Contact: Linda Sensenig
345 Douglass Street
Wyomissing, PA 19610
Phone: 610-375-9062

Bucks-Mont Astronomical Association
E-mail: bmaa@ixc.net
Internet: astro4.ast.vill.edu/
    bmaatop.htm
Contact: Stephen Bryant
607 Manor Drive
Dublin, PA 18917
Phone: 215-249-1109
E-mail: sbryant@sungard.com

Delaware Astronomical Society
Contact: Emil J. Volcheck
1046 General Allen Ln
West Chester, PA 19382-8030
Phone: 610-388-1581
E-mail: 74425.405@compuserve.com

Delaware Valley Amateur Astronomical Society
Internet: www.libertynet.org/~dvaa
Contact: Dr. Paul Perlmutter
83 Llanfair Circle
Ardmore, PA 19003-3342
Phone: 610-642-9142
E-mail: mnbv64a@prodigy.com

Lehigh Valley Amateur Astronomical Society
E-mail: lvaas@lvaas.org
Internet: www.lvaas.org
Contact: Robert Mohr
Box 368
7535 Tilghman Street
Fogelsville, PA 18051
Phone: 610-398-7295

Sir Isaac Newton Astronomical Society
Contact: William Kearney
434 Greeves Street
Kane, PA 16735
Phone: 814-837-7205

York County Parks Astronomical Society
Contact: David M. Dewey
5 Hilltop Drive
New Freedom, PA 17349
Phone: 717-235-3779

Puerto Rico

Sociedad de Astronomia de Puerto Rico
Contact: Ernesto E. Santiago-Jordan
34th Street, P-2
Urb. Bairoa
Caguas, PR 00725
Phone: 809-743-8839

South Carolina

Carolina Skygazers Astronomy Club
Museum of York County
4621 Mt. Gallant Road
Rock Hill, SC 29732
Phone: 803-329-2121
Contact: Barbara Reynolds
2067 Marquesas Avenue
Tega Cay, SC 29715

Low Country Stargazers
Contact: Ken Watson
1022 Margaret Drive
Ladson, SC 29456
Phone: 803-873-9653

Tennessee

Barnard Astronomical Society
Contact: Frank R. Helms
838 Belvoir Hills Drive
Chattanooga, TN 37412-2016
Phone: 423-629-6094

Barnard-Seyfert Astronomical Society
c/o A.J. Dyer Observatory
1000 Oman Drive
Brentwood, TN 37027
Contact: Michael G. Benson
Phone: 615-883-6571
E-mail: Ocentaurus@aol.com

Bays Mountain Astronomy Club
853 Bays Mountain Park Road
Kingsport, TN 37660
Phone: 423-229-9447
Contact: Mike Chesman
605 Rambling Road
Kingsport, TN 37663
Phone: 615-229-9447
E-mail: baysmtn@tricon.net

Smoky Mountain Astronomical Society
Contact: Barry Cain
430 Broome Road
Knoxville, TN 37909
Phone: 423-531-7181

Society of Low-Energy Observers
Contact: William J. Cupo, Jr.
3260 Spottswood Avenue
Memphis, TN 38111
Phone: 901-366-4201

Texas

Amarillo Astronomy Club
E-mail:rdshep@arn.net
Contact: Don Chrysler
#13 Quadrille St.
Amarillo, Texas 79106
Phone: 806-359-8988

American Association of Amateur Astronomers
3131 Custer Road, Suite 175/175
Plano, TX 75075
E-mail: aaaa@corvus.com
Contact: John Wagoner
1409 Sequoia
Plano, TX 75023
Phone: 214-422-1886
E-mail: john.wagoner@corvus.com

Astronomical Society of East Texas
Contact: Larry D. Scott
10657 C.R. 214
Tyler, TX 75707-9517
Phone: 903-566-5317
E-mail: 71045.213@compuserve.com

Astronomical Society of North Texas
Route 1, Box 323A5
Bonham, TX 75418
E-mail: pdturner@juno.com
Internet: www.netexas.net/pturner
Contact: Mr. Jim McGlynn
125 Watkins Road
Sherman, TX 75090
Phone: 903-870-0080

Astronomical Society of South East Texas
Contact: Dave Deming
P. O. Box 461
Mauriceville, TX 77626
Phone: 409-745-1366

Big Bend Astronomical Society, Inc.
HC 65, Box 14
Alpine, TX 79830
Contact: Eileen Conner
P. O. Box 666
Marfa, TX 79843
Phone: 915-729-3541
E-mail: EilCon@aol.com

Crossroads Astronomy Club
P. O. Box 5099
Victoria, TX 77903
Contact: Mr. K. B. Hallmark
P. O. Box 180
Nursery, TX 77976
Phone: 512-578-7942

Fort Bend Astronomy Club
Internet: rampages.onramp.net/~binder/
Contact: Tracy Knauss
1112 Wayne Drive, #5
Angleton, TX 77515
Phone: 409-345-2521

Fort Worth Astronomical Society
P. O. Box 161715
Fort Worth, TX 76161-1715
Internet: www.flash.net/~rickc/
    fwas.html.
Contact: Jay Hornsby
3515 Ashley St.
Arlington, TX 76016
Phone: 451-8948

Houston Astronomical Society
Internet: spacsun.rice.edu/~has.
Contact: Steve Goldberg
5115 Stillbrooke
Houston, TX 77035
Phone: 713-721-5077
E-mail: goldberg@sccsi.com

McLennan County Astronomy Club
Phone: 817-750-5390
E-mail: 1bdb1974@tstc.edu
Internet: www.tstc.edu/~1bdb1974/
    mcac.htm
Contact: Bernard F. Ott, Jr.
612 Kipling Drive
Waco, TX 76710
Phone: 817-776-5353
E-mail: b3c@juno.com

San Angelo Amateur Astronomical Assoc.
P. O. Box 60391
San Angelo, TX 76906
Contact: Jeff Foreman
P. O. Box 60391
San Angelo, TX 76906
Phone: 915-944-3928

San Antonio Astronomical Association
4438 Bikini Drive
San Antonio, TX 78218
E-mail: rondawes@aol.com
    or 73766.470@compuserve.com
Contact: Ed Sarratt
463 Sandalwood
San Antonio, TX 78216
Phone: 210-828-3271

South Plains Astronomy Club
Internet: www.math.ttu.edu/
    ~wlewis/astro.html
Contact: Wayne Lewis
4803 76th Street
Lubbock, TX 79424-2148
Phone: 806-794-8766
E-mail: wlewis@math.ttu.edu

Texas Astronomical Society of Dallas
P. O. Box 25162 Preston Station
Dallas, TX 75225
E-mail: tasclub@airmail.net
Internet: web2.airmail.net/tasclub
Contact: James W. Bandy
6932 Quarterway Dr.
Dallas, TX 75248-5547
Phone: 972-490-6483
E-mail: jbandy@airmail.net

West Texas Astronomers
Blakemore Planetarium
Museum of the Southwest
1705 W. Missouri
Midland, TX 79701-6516
Phone: 915-689-6375
Contact: Bernard Lucas
3100 Metz Drive
Midland, TX 79705
Phone: 915-689-6375

Virginia

Back Bay Amateur Astronomers
Contact: Glendon L. Howell
2808 Flag Road
Chespeake, VA 23323-2102
Phone: 757-485-4242

Blue Ridge Skywatch
Astrophotography Society
Contact: S. G. Marshall
1611 Cliffview Drive
Salem, VA 24153
Phone: 703-387-3230

Bristol Astronomy Club
Contact: Ken Childress
4559 Reedy Creek Road
Bristol, VA 24202
Phone: 540-466-8404

Northern Virginia Astronomy Club
Internet: astro.gmu.edu/~novac
Contact: Bill Jensen
7405 Ridge Oak Ct.
Springfield, VA 22153
Phone: 703-866-1380
E-mail: william.jensen@tcs.wap.org

Richmond Astronomical Society
E-mail: bwilson@pen.k12.va.us
Contact: Henry W. Stockmar, III
2218 Martin Street
Richmond, VA 23228
Phone: 804-262-9151

Roanoke Valley
Astronomical Society
E-mail: brossie@intrlink.com
Contact: Liana Arias de Velasco
2801 Brandon Avenue
Roanoke, Virginia 24015
Phone: 540-985-3163
E-mail:
    LAV@LYNXOS3.SALEM.GE.COM

Washington

Everett Astronomical Society
P. O. Box 8012
Everett, WA 98201
Contact: Joanne Green
5307 30th Avenue NE
Seattle, WA 98105
Phone: 206-524-2006
E-mail: joagreen@aol.com

Olympic Astronomical Society
Contact: Roger Miller
147 NE Watson Court
Bremerton, WA 98311
Phone: 360-692-6300

SW Washington
Astronomical Society
Contact: Glenn Varano
2136 Woodcrest Drive
Olympia, WA 98501
Phone: 360-943-1140

Seattle Astronomical Society
E-mail: xx004@scn.org
Internet: www.scn.org/ip/sastro/
Contact: Loren Busch
8340 17th NW
Seattle, WA 98117
Phone: 206-782-3496

Spokane Astronomical Society
P. O. Box 8114
Spokane, WA 99203
E-mail: dan@runaway.net
Internet: www.runway.net/a/sas/
Contact: Chuck Cheselka
1808 W. Knox
Spokane, WA 99205-4144
Phone: 509-328-9468

Tacoma Astronomical Society
Contact: Carl Tankersley
1501-A 128th E.
Tacoma, WA 98445
Phone: 206-531-0858
E-mail: carltank@aol.com

Titchenal Canyon Group
10 N. Western
Wenatchee, WA 98802
Contact: Vladimir Steblina
209 N. Nancy
East Wenatchee, WA 98802
Phone: 509-884-4491

Tri-City Astronomy Club
P. O. Box 651
Richland, WA 99352
Contact: Barbara Rittmann
5001 W. Skagit Ave
Kennewick, WA 99336
Phone: 509-783-1845

Whatcom Association of Celestial Observers (WACO)
Internet: www.wwu.edu/~skywise/
    waco.html
Contact: Kreig McBride
402 East Horton Road
Bellingham, WA 98226
E-mail: kmcbride@kali.nas.com

Yakima Valley Astronomy Club
E-mail: mknight@wolfenet.com
Contact: Dave Trick
611 South 17th Avenue
Yakima, WA 98902
Phone: 509 457-2785
E-mail: dtrick@wolfenet.com

Wisconsin

Chippewa Valley Astronomical Society
S1 County Road K
Fall Creek, WI 54742
E-mail: elliott@uwec.edu
Internet: www.phys.uwec.edu/cvas/
Contact: Ray Forsgren
4856 Inlet Dr.
Chippewa Falls, WI 54729-9119
Phone: 715-723-8489
E-mail: starman@overlord.edp.net

LaCrosse Area Astronomical Society
Contact: Robert Allen
Physics Dept, Cowley Hall
University of Wisconsin
LaCrosse, WI 54601
Phone: 608-785-8669
E-mail: robert_allen@uwlax.edu

Milwaukee Astronomical Society
MAS Observatory
18850 W. Observatory Road
New Berlin, WI
Phone: 414-542-9071
Contact: Julie Frey
11040 West Mienecke Avenue, #4
Wauwatosa, WI 53226
Phone: 414-456-0864
E-mail: juliekfrey@aol.com

Neville Public Museum Astronomical Society

Contact: Ronald Parmentier
161 Rosemont Drive
Green Bay, WI 54301
Phone: 414-336-5878

Northeast Wisconsin Stargazers

Contact: Terry Becker
514 Union Avenue, Apt. E
Oshkosh, WI 54901
Phone: 414-426-2286
E-mail: tlb@vbe.com

Northern Cross Science Foundation

1219 12th Avenue
Grafton, WI 53024-1923
E-mail: matthies@omnifest.uwm.edu
Contact: Kevin Bert
2292 Ridgewood Road
Grafton, WI 53024-9546
Phone: 414-375-2239

Racine Astronomical Society

P. O. Box 085694
Racine, WI 53408
E-mail: rasastro@wi.net
Internet: http://www.iwc.net/~rasastro/
Contact: Richard Wend
37754 N. Lake Vista Terrace
Spring Grove, IL 60081
Phone: 847-587-0492

Rock Valley Astronomers

Contact: Carl Balke
2220 E. Ridge Road
Beloit, WI 53511
Phone: 608-362-0432

Sheboygan Astronomical Society

E-mail: dvminsel@lakefield.net
Internet: bratshb.uwc.edu/~sas/
Contact: Mr. Kurt Petersen
313 Main Street
Sheboygan Falls, WI 53085
Phone: 414-467-2257
E-mail: LoonP@aol.com

Wehr Astronomical Society

Wehr Nature Center
9701 W. College Avenue
Franklin, WI 53132
Contact: Charlotte Nelson
4151 N. 98th Street
Wauwatosa, WI 53222-1422
Phone: 414-466-2081

Appendix B
National Space Society (NSS) Chapters

Alabama

Huntsville Alabama L5 Society
1019A Old Monrovia Road, Suite 168
Huntsville, AL 35806
Chapter Contact: Gregory H. Allison
Phone: 205/859-5538
E-Mail: hal5@iquest.com
Newsletter:
    Southeastern Space Supporter
Newsletter Editor: Ronnie Lajoie
Newsletter Phone: 205/461-3064
Web Page: http://iquest.com/~hal5/

Arkansas

First Arkansas Chapter of NSS
c/o Gerhard Langguth
243 White Road
Russellville, AR 72801
Chapter Contact: Gerhard Langguth
Phone: 501/967-4531
E-Mail: glanggut@cswnet.com

Arizona

Tucson Space Society
c/o David Brandt-Erichsen
5100 N. Moonstone Drive
Tucson, AZ 85747
Phone: 520/749-2247
E-Mail: davidbe@azstarnet.com
Web Page: http://www.azstarnet.com/
    ~dhfred/tsshome.html

California

Golden Gate Space Frontier Society
4009 Everett Avenue
Oakland, CA 94602
Chapter Contact: Ray Miller
Phone: 510/530-1971
Newsletter: *Spacefaring Gazette*
Newsletter Editor: Jaki Jepson
Newsletter E-Mail: raygmiller@aol.com

NSS Western Spaceport Chapter
c/o James Spellman, Jr.
4617 Oak Lane, Mtn. Mesa
Lake Isabella, CA 93240-9713
Chapter Contact: Jim Spellman
Phone: 619/379-2503
E-Mail: wspaceport@aol.com
Newsletter: *Western Space Report*
Newsletter Editor: Jim Spellman

OASIS
PO Box 1231
Redondo Beach, CA 90278
Chapter Contact: Mark Holthaus
Phone: 310/594-0272
E-Mail: oasis-leaders@netcom.com
    or. normanc523@aol.com
Newsletter: *Odyssey*
Newsletter Editor: Norm Cook
Web Page: http://members.aol.com/
    OASISnss

Orange County Space Society
25671 Le Parc, #89
Lake Forest, CA 92630
Chapter Contact: Larry Evans
Phone: 714/770-0702
E-Mail: Mach25Comm@aol.com
Newsletter: *OCSS News*
Newsletter Editor: Phil Turek

Sacramento L5 Society
c/o Robert Compton
3945 Grey Livery Way
Sacramento, CA 95843
Chapter Contact: Robert Compton
Phone: 916/344-3290
E-Mail: 74274.2074@compuserve.com

San Diego L5-A Chapter of NSS
PO Box 4636
San Diego, CA 92164
Chapter Contact: Gregory Nemitz
Phone: 619/525-3820
E-Mail: greg@beefjerky.com
Newsletter: *VOX INTER ASTRA*
Newsletter Editor: Susan Abernathy

Silicon Valley Space Society
PO Box 60194
Sunnyvale, CA 94088
Chapter Contact: Kurt Bohan
Phone: 510/783-5903
E-Mail: kbohan@kaylor-kit.com
Newsletter: *Silicon Valley News*
Newsletter Editor: Kurt Bohan

Colorado

Colorado Springs Space Society
c/o Theresa Holmes
3303 N. Hancock Ave., #86
Colorado Springs, CO 80907
Chapter Contact: Theresa Holmes
Phone: 719/520-0469
E-Mail: sstohot@aol.com

Front Range L5 Society
PO Box 291
7450 W 52nd Avenue, Suite M
Arvada, CO 80002
Chapter Contact: Neil Ducher
Phone: 303/455-0508
E-Mail: yinmore@aol.com

Mile High L5 Society
c/o Mark Schloesslin
6937 E. Briarwood Circle
Englewood, CO 80112
Chapter Contact: Mark E. Schloesslin
Phone: 303/779-5692

Northern Colorado Space Society
c/o Cheryl Whiston
8638 WCR 26
Ft. Lupton, CO 80621-9721
Chapter Contact: Cheryl Whiston
Phone: 303/785-2580

Connecticut

Central Connecticut Space Development
c/o Victor Morris
146 West Street
Cromwell, CT 06416
Chapter Contact: Victor Morris
Phone: 203/635-6090

Connecticut Chapter of NSS
PO Box 230628
Hartford, CT 06123-0628
Chapter Contact: William Bourn
Phone: 203/633-3154
E-Mail: tncmaxq@aol.com

District of Columbia

National Capital Space Society
c/o Bob Kozon
233 Kentucky Ave., SE
Washington, DC 20003
Chapter Contact: Bob Kozon
Phone: 202/543-2137
E-Mail: kozon@gsfc.nasa.gov

Florida

Embry-Riddle Aerospace Society
c/o Student Activities Office
600 S. Clyde Morris Blvd.
Daytona Beach, FL 32114-3900
Chapter Contact: Dr. Lance Erickson
Phone: 904/760-6651

Metro Orlando Space Society
PO Box 1829
Orlando, FL 32802-1829
Chapter Contact: Michael J. Gilbrook
Phone: 407/876-6731
E-Mail: 72237.3611@compuserve.com
Newsletter:
Central Florida Space Activist Update
Newsletter Editor: Ed Scull

NSS of the Palm Beaches
Dusty McGee
14369 Stirrup Lane
Wellington, FL 33414-8220
Phone: 561/795-1943
E-Mail: macdust@flinet.com
Web Page: http://www.joek.com/
   nsspb.html

Tallahassee Per Spatium
c/o Sheridan Layman
1491 Twin Lakes Circle
Tallahassee, FL 32311
Phone: 904/877-6922
E-Mail: sabo@magnet.fsu.edu

Georgia

Central Savannah River Area NSS
4578 Brandermill Court
Evans, GA 30809
Chapter Contact: Henry Quinn
Phone: 706/863-7641
E-Mail: hjquinn@csranet.com

NSS Atlanta
2105 Hunter Ridge Lane
Norcross, GA 30092
Chapter Contact: Ben King
Phone: 770/446-8212
E-Mail: benking@paradise.net
Web Page: http://www.netvisibility.com/
   nssatl

NSS Northeast Georgia Chapter
c/o Greg Rucker
635 Huntington Road, #402
Athens, GA 30606
Chapter Contact: Greg Rucker
Phone: 706/543-1590
E-Mail: ggrucker@music.cc.uga.edu
Web Page: http://www.redhouse.com/
   space

Hawaii

Hawaii Space Society
PO Box 61206
Honolulu, HI 96839-1206
Chapter Contact: Gregory A. Smith
Phone: 808/956-3158
E-Mail: gsmith@pgd.hawaii.edu
Web Page: http://apollo-society.org/
   apollo/hss.html

Illinois

Chicago Society for Space Studies
PO Box 1454
North Riverside, IL 60546
Chapter Contact: Lawrence Boyle
Phone: 708/788-1336
Newsletter: *Spacewatch*
Newsletter Editor: Jim Plaxco
Newsletter Phone: 708/924-7122
Newsletter E-mail:
72500.1724@compuserve.com

Chicago Space Frontier L5 Society
c/o Bill Higgins
MS 355, Fermilab Box 500
Batavia, IL 60510
Chapter Contact: Bill Higgins
Phone: 708/393-6817

Illinois Space Development Society
Dept. of Aero-Astro Engineering
314 Talbot Laboratory
Urbana, IL 61801
Chapter Contact: Ross Beyer
Phone: 217/328-5401
E-Mail: rossb@astro.uiuc.edu
Web Page: http://stimpy.cen.uiuc.edu/
soc/isds/

Illinois North Shore NSS
c/o Jeffrey Liss
1364 Edgewood Lane
Winnetka, IL 60903
Chapter Contact: Jeff Liss
Phone: 708/446-8343
E-Mail: jgljgl@aol.com

Rock Valley Space Frontier Society
c/o Bryce Johnson
1441 Greenwood Avenue
Rockford, IL 61107
Phone: 815/967-0490
E-Mail: arr3@aol.com

Zero G Destiny
PO Box 971
Homewood, IL 60430-0971
Chapter Contact: Jerry Smith
Phone: 708/758-3557
E-Mail: jerrys3971@aol.com
Newsletter: *The Rail*
Newsletter Editor: Tim Cash

Indiana

Grissom Space Society
c/o Randall Porter
306 S. Gale St.
Indianapolis, IN 46201-4442
Chapter Contact: Randall Porter
Phone: 317/274-1535
Newsletter: *Out of the Well*
Newsletter E-Mail:
wrporter@indyvax.iupui.edu

Iowa

Iowa State Space Society
PO Box 2465
Ames, IA 50010
Chapter Contact: Andy Stevenson
Phone: 515/296-9556
E-Mail: issscabinet@iastate.edu
Newsletter: *Space Insider*

Kansas

Wichita Chapter of NSS
c/o Glen Burdue
741 N. Clara
Wichita, KS 67212-2661
Chapter Contact: Glen Burdue
Phone: 316/943-8880

Kentucky

Kentucky Chapter of NSS
c/o Harry Reed
163 Harrison Rd.
Benton, KY 42025
Chapter Contact: Harry Reed
Phone: 502/527-1087

Maryland

Baltimore Metro Chapter of NSS
235 Wakely Terrace
Bel Air, MD 21014
Chapter Contact: Dale S. Arnold, Jr.
Phone: 410/879-3602

Massachusetts

**National Space Society,
Boston Chapter**
c/o Roxanne Warniers
5 Driftwood Lane
Acton, MA 01720
Chapter Contact: Andrew Lepage
E-Mail: bam@draper.com
    or: lepage@bur.visidyne.com
Web Page: http://www.seds.org/
    spaceviews/boston.html
Newsletter: *SpaceViews*
Newsletter Editor: Jeff Foust
Newsletter E-Mail: jeff@astron.mit.edu

Michigan

Ann Arbor Space Society
PO Box 130118
Ann Arbor, MI 48113-0118
Chapter Contact: William Bogen
Phone: 313/769-5223
E-mail: wpb@iti.org

NSS Southwest Michigan Chapter
c/o Lon Grover
13613 East L Ave.
Galesburg, MI 49053
Chapter Contact: Lon Grover
Phone: 616/746-5268
E-Mail: louann.grover@wmich.edu

Minnesota

Minnesota Space Frontier Society
PO Box 581939
Minneapolis, MN 55458-1939
Chapter Contact: Jeff Root
Phone: 612/375-1539
E-Mail: mn.sfs@mnspace.mn.org
Newsletter: *Downrange*
Newsletter Editor: Ben Huset
Web Page: http://www.skypoint.com/
    subscribers/benhuset/mnsfsinf.html

Missouri

Heart of America Chapter NSS
c/o Robert Stout
2918 South Sterling Ave.
Independence, MO 64052-3063
Chapter Contact: Robert Stout
Phone: 816/836-2766
E-Mail: rstout@aol.com

St. Louis Space Frontier
PO Box 342
Hazelwood, MO 63042
Chapter Contact: Tom Becker
E-Mail: ransompw@aol.com
Newsletter: *The New Frontier*
Newsletter Editor: Tom Schultz
Newsletter Phone: 314/925-5790

New Jersey

NSS Northern New Jersey Chapter
c/o Greg Zsidisin
PO Box 71
Maplewood, NJ, 07040
Chapter Contact: Greg Zsidisin
Phone: 201/762-3270
E-Mail: nssnj@aol.com
Newsletter: *Moon Miners Manifesto*
Newsletter Editor: Peter Kokh

New Mexico

Albuquerque NSS
c/o Ken Newman
4917 Tally Ho NW
Albuquerque, NM 87114
Chapter Contact: Ken Newman
Phone: 505/898-7796
E-mail: knewman@usa.net
Newsletter: *Albuquerque NSS Update*
Newsletter Editor: Curtis Smith

New York

New Frontier Society
117 Kirklees Rd.
Pittsford, NY 14534
Chapter Contact: Carl Elsbree
Phone: 716/427-4207

Rennsselaer Space Society
RPI Student Union
PO Box 86
Troy, NY 12180
Chapter Contact: Christine Tyrell
E-Mail: tyrelc@rpi.edu, hylanj@rpi.edu

Space Frontier Society of New York City
PO Box 1985
New York, NY 10163
Chapter Contact: Greg Zsidisin
Phone: 212/SAM-8000
E-Mail: nssnyc@aol.com
Newsletter: *Inside NSS*
Newsletter Editor: Jeff Liss
Web Page: http://www.mne.com/sfs/

Suffolk Challengers for Space
c/o Sue Lorraine Lavorata
182 Millard Avenue
West Babylon, NY 11704
Chapter Contact: Sue Lorraine Lavorata
Phone: 516/321-0964
E-Mail: slavorat@help.her.sunysb.edu

North Carolina

Triangle Space Society
c/o Steve Nixon
4213 Bertram Dr. #11
Raleigh, NC 27604
Chapter Contact: Steve Nixon
Phone: 919/954-9615
E-Mail: rocketm@nando.net

Ohio

Cuyahoga Valley Space Society
c/o George F. Cooper, III
3433 North Avenue
Parma, OH 44134
Chapter Contact: George F. Cooper, III
Phone: 216/749-0017
Newsletter: *Island One Journal*
E-Mail: hi743@cleveland.freenet.edu

Oklahoma

Oklahoma Space Alliance NSS
c/o Claire Stephens
1206 Classen Blvd.
Norman, OK 73071
Chapter Contact: Claire Stephens
Phone: 405/329-4326
E-Mail:
    claire@ess-lan.studies.uoknor.edu
Web Page: http://www.telepath.com/
    osa/index.html

Oregon

Deschutes Space Frontier Society
c/o Mike Guidero
65128 Hunnell Rd.
Bend, OR 97701
Chapter Contact: Mike Guidero
Phone: 503/388-3781
E-Mail: rrrspace@aol.com

NSS Southern Oregon Chapter
520 S. Oregon St.
PO Box 445
Jacksonville, OR 97530
Chapter Contact: Brian Lundquist
Phone: 503/899-8504
E-Mail: wind12@cdsnet.net

Oregon L5
PO Box 42467
Portland, OR 97242-0467
Chapter Contact: Tom Billings
Phone: 503/655-6189
E-Mail: pioneertom@aol.com
Newsletter: *Starseed*
Newsletter Editor: Bryce Walden

Oregon State University L5
3420 NW Elmwood Dr.
Corvalis, OR 97339
Chapter Contact: Gary Oliver
Phone: 503/757-0634
E-Mail: go@ao.com

Pennsylvania

Beaver Valley Space Education
2509 Steffen Hill Road
Beaver Falls, PA 15010
Chapter Contact: William Cress
Phone: 412/846-7808

Northeastern Pennsylvania Space Foundation
c/o Daniel A. Freedman
646 Ford Avenue
Kingston, PA 18704
Chapter Contact: Michelle Girvan
Phone: 717/288-7075
E-Mail: girvanm@wilkes1.wilkes.edu
Newsletter: *North Coast's Trajectory*

NSS North Coast Chapter
c/o Jim Carnes
2925 Hemlock Dr.
Erie, PA 16506
Chapter Contact: Jim Carnes
Phone: 814/833-6095
E-Mail: 74364.37@compuserve.com
Newsletter: *Trajectory*
Newsletter Editor: Jim Carnes

Philadelphia Area Space Alliance
PO Box 1715
Philadelphia, PA 19105
Chapter Contact: Michelle Baker
Phone: 609/561-8867
E-Mail:
    michelle_baker@ccgate.ueci.com
or: hainesjb@netaxs.com
Newsletter: *Moon Miners Manifesto*
Newsletter Editor: Peter Kokh
Newsletter Phone: 414/342-0705
Newsletter E-Mail:
    kokhmmm@aol.com

South Dakota

South Dakota Space Society
c/o John Sullivan
217 S. Summit #2
Sioux Falls, SD 57104
Chapter Contact: John Sullivan
Phone: 605/332-6745
E-Mail: bnelson@tth.net

Tennessee

East Tennessee Space Society
c/o Virden Spicer, Jr.
13 Converse Lane
Oak Ridge, TN 37830
Chapter Contact: Virden Spicer, Jr.
Phone: 615/481-0742
Newsletter: *Moon Miners Manifesto*
Newsletter Editor: Peter Kokh
Newsletter Phone: 414/342-0705
Newsletter E-Mail: kokhmmm@aol.com

Memphis Space Society
c/o Robert Farris
3607 Kipling Avenue
Memphis, TN 38128
Chapter Contact: Robert Farris
Phone: 901/382-3135
E-Mail:
    73014.1466@compuserve.com

Middle Tennessee Space Society
c/o Chuck Shlemm
508 Beechgrove Way
Burns, TN 37029
Chapter Contact: Chuck Shlemm
Phone: 615/441-1024
Newsletter: *The Rocket*
Newsletter Editor: Scott Pearson

Willy Ley Space Society
c/o Lee Cummings
6636 Shallowford Rd.
Chattanooga, TN 37421
Chapter Contact: Lee Cummings
Phone: 615/855-0303

Texas

Austin Space Frontier Society
c/o John Strickland
12717 Bullick Hollow Road
Austin, TX 78726
Chapter Contact: John Strickland
Phone: 512/258-8998
E-mail: jkstrick@io.com

Clear Lake Area Chapter of NSS
c/o Marianne J. Dyson
15443 Runswick Drive
Houston, TX 77062-3310
Chapter Contact: Murray G. Clark
Phone: 281/367-2227
E-mail: MClark637@aol.com

NSS of North Texas
PO Box 1671
Arlington, TX 76004-1671
Chapter Contact: Carol Johnson
Phone: 214/937-3587
E-Mail:
    73302.3202@compuserve.com
Newsletter: *The Mid Cities Spacecraft*
Newsletter Editor: Curtis Kling

San Antonio Space Society
Joe B. Redfield c/o SwRI
6220 Culebra
San Antonio, TX 78228
Chapter Contact: Joe B. Redfield
Phone: 210/522-3729
E-Mail: jredfield@swri.edu

Southeast Texas Space Initiative
7070 Shanahan
Beaumont, TX 77706
Chapter Contact: Dale Parish
Phone: 409/745-3899
E-Mail: wjwv67a@prodigy.comu
Newsletter: *The Initiator*
Newsletter Editor: Dale Parish

Virginia

DC L5
PO Box 16630
Arlington, VA 22215-1630
Chapter Contact: Bennett Rutledge
Phone: 703/780-0434
E-Mail: rutledges@delphi.com

Langley NSS
7716 Gloucester Ave.
Norfolk, VA 23505
Chapter Contact: David Hahn
Phone: 804/423-8143
E-Mail: ggrahn@aol.com
Newsletter: *LaNSS News*

Washington

NSS Seattle
Vince Creisler
13241 S.E. 261st Street
Kent, WA 98042
Phone: 206/630-4735
E-Mail: Vincelc@worldnet.att.net

Wisconsin

Lunar Reclamation Society, Inc.
PO Box 2102
Milwaukee, WI 53201-2102
Chapter Contact: Mark Kaehny
Phone: 414/572-2054
Newsletter: *Moon Miners Manifesto*
Newsletter Editor: Peter Kokh
Newsletter Phone: 414/342-0705
Newsletter E-Mail:
    kokhmmm@execpc.com

Sheboygan Space Society
728 Center St.
Kiel, WI 53042-1034
Chapter Contact: Wil Foerster
Phone: 414/894-2376
E-Mail: haralds971@aol.com

Australia

Adelaide Space Frontier Society
18 Charmaine Ave.
Para Vista, SA 5093
AUSTRALIA
Chapter Contact: Stewart Clarke
Phone: +61 08 322-6979

Canberra Space Frontier Society
PO Box 7048
Sydney, NSW 2001
AUSTRALIA
Chapter Contact: Martin Thorne
Phone: +61 02 560-0060

Hobart Space Frontier Society
PO Box 7048
Sydney, NSW 2001
AUSTRALIA
Chapter Contact: Martin Thorne
Phone: +61 02 560-0060

Melbourne Space Frontier Society
c/o SPACElink Consulting
PO Box 2375
Brighton North, Vic 3186
AUSTRALIA
Chapter Contact: Frank Papa
Phone: (+61 3) 9592 0935
Fax: (+61 3) 9592 0813
E-Mail: spacelink_consulting
@compuserve.com

Newcastle Space Frontier Society
PO Box 1150
Newcastle, NSW 2300
AUSTRALIA
Chapter Contact: Glen Eastlake
Phone: +61 049 63-5037
E-Mail: dis1@castle.net.au

Perth Space Frontier Society
PO Box 7048
Sydney, NSW 2001
AUSTRALIA
Chapter Contact: Martin Thorne
Phone: +61 02 560-0060

Queensland Space Frontier Society
c/o Allen Reynolds
PO Box 419
Nundah, QLD 4012
AUSTRALIA
Chapter Contact: Allen Reynolds
Phone: +61 07-266-8839

Sydney Space Frontier Society
PO Box 7048
Sydney, NSW 2001
AUSTRALIA
Chapter Contact: Martin Thorne
Phone: +61 02 560-0060

**Western Sydney Space
Frontier Society**
c/o Ralph Buttigieg
PO Box 1083
Auburn, NSW 2144
AUSTRALIA
Chapter Contact: Ralph Buttigieg
Phone: +61 02 635-6797
E-Mail: rbuttigieg
@vulcans.caamora.com.au

Canada

**Niagara Peninsula Space Frontier
Society**
c/o Raymond Merrick
PO Box 172
Thorold, Ontario L2V 3Y9
CANADA
Chapter Contact: Raymond Merrick
Phone: 905/984-4086

Mexico

Sociedad Espacial Mexicana, A.C.
c/o Jesus Raygoza B.
Apartado Postal 5-75
Guadalajara, JALISCO 45042
MEXICO
Chapter Contact: Jesus Raygoza B.
Phone: (3) 647-5710

United Kingdom

**Brunel School For
Space Education NSS**
c/o Jim Potter
Brunel University
Uxbridge UB8 3PH
UNITED KINGDOM
Chapter Contact: Jim Potter
Phone: +44 895 271206

Special Interest Chapters

Molecular Manufacturing Shortcut Group
c/o Tom McKendree
8381 Castilian Dr.
Huntington Beach, CA 92646
Chapter Contact: Tom McKendree
Phone: 714/374-2081
E-Mail: ttf@dsg066.nad.ford.com
or: tmckendree@msmail3.hac.com
Newsletter: *The Assembler*
Newsletter Editor: Ti Toth-Fejel
Web Page: http://www.islandone.org/
    MMSG/

NSS Environment Chapter
c/o David Anderman
7584 Rush River Dr. #29
Sacramento, CA 95381
Chapter Contact: Forrest Schultz
Phone: 404/583-3258
E-Mail: david@cwnet.com
Newsletter:
    Space and the Environment News
Newsletter Editor: Thomas Cleveland

Miscellaneous Contacts

Chapters Coordinator
Craig Ward
1914 Condon Ave.
Redondo Beach, CA 90278-3403
Phone: 310/371-7015
E-Mail: cew@acm.org

U.S. Chapters Coordinator
Shirley Smith
110 N. Wabash Ave.
Glenwood, IL 60425
Phone: 708/758-3557
E-Mail: Jerrys3971@aol.com

International Chapters Coordinator
c/o Kirby IKIN
GPO BOX 7048
Sydney, NSW 2001
AUSTRALIA
Phone: +61 2-228-1327
E-Mail: angiokdi@ibmmail.com

NSS Chapters Assembly
c/o Ronnie M. Lajoie, Chair
162 Kirby Lane
Madison, AL 35758
Phone: 205/461-3064 (day)
Phone: 205/721-1083 (evening)
FAX: 205/461-2551
E-Mail: hal5@iquest.com
Focus: Helping NSS chapters thrive
Newsletter: *Inside NSS*
Newsletter Editor: Jeff Liss
Newsletter Phone: 708/446-8343
Newsletter E-Mail: jgljgl@aol.com
Web Page: http://iquest.com/~hal5/
    assembly/

Appendix C

NASA Teacher Resources

NASA maintains an extensive network of resources to assist educators and students at all levels. These resources include the tri-annual publication, *Educational Horizons*, for educators, on-line services like *Spacelink* and *Quest*, NASA Television, the Education Satellite Videoconference Series for Teachers, the Central Operation of Resources for Educators (CORE), and the Teacher Resource Center Network (TRCN), which includes Teacher Resource Centers and Regional Teacher Resource Centers across the country. The following pages provide contact information for these programs, including a state-by-state listing of Regional Teacher Resource Centers.

NASA's Educational Horizons Newsletter

Educational Horizons
NASA
Education Division
Mail Code FE
Washington, DC 20546-0001
Internet: spacelink.msfc.nasa.gov

Published in April, September and December, *Educational Horizons* is NASA's newsletter for educators. The publication provides news on NASA Shuttle missions, updates on NASA space science and aeronautics research, and information on NASA education resources, including the latest educational publications available from the Agency.

To subscribe, or to request additional information, contact the Teacher Resource Center or Regional Teacher Resource Center nearest you. If accessing *Educational Horizons* electronically via the Internet, follow directory path Educational.Services, Educational.Publications, Educational.Horizons.Newsletter.

On-line Services

NASA Spacelink
Education Programs Office
Mail Code CL 01
NASA Marshall Space Flight Center
Huntsville, AL 35812-0001
Phone: (205) 961-1225
E-mail:
  comments@spacelink.msfc.nasa.gov
Internet: spacelink.msfc.nasa.gov

Spacelink is an electronic resource which can be accessed on the Internet at http://spacelink.msfc.nasa.gov. Designed to provide current educational information to teachers, faculty, and students, Spacelink offers text files, software, and graphics related to the aeronautics and space program. The site contains both historical information on the space program, as well as current status reports on NASA projects, news releases, and more.

Quest
E-mail: info@quest.arc.nasa.gov
Internet: quest.arc.nasa.gov

Quest is NASA's K-12 Internet Initiative, a resource the Agency developed specifically for the educational community. Teachers can access information about educational grants, interact with other schools, and explore other NASA educational resources. Quest also features the "Sharing NASA" initiative, which allows students and educators to communicate with NASA scientists and researchers.

NASA Television (NTV)

NASA Headquarters
Code P-2
NASA TV
Washington, CD 20546-0001
Phone: (202) 358-3572
Internet: www.hq.nasa.gov/office/pao/ntv.html

NASA Televison provides live mission coverage, News Video Files, NASA History Files, and Education Files at regularly scheduled times, Monday through Friday. NASA Television can be access through some cable systems, or via satellite dish. The program is transmitted on Spacenet 2 (a C-band satellite) on transponder 5, channel 9, 69 degrees west with horizontal polarization, frequency 3880 MHZ, audio on 6.8 MHZ.

Education Satellite Videoconference Series for Teachers

Videoconference Coordinator
NASA Teaching From Space Program
308 CITD Room A
Oklahoma State University
Stillwater, OK 74078-8089
Phone: (405) 744-6784
Fax: (405) 744-6785
E-mail: edge@aesp.nasa.okstate.edu

Videoconference series a live and interactive, allowing educators to call the studio to ask questions. Topics include presentations by a NASA astronaut, project scientist, or program administrator, updates on current NASA educational products, services, and programs, demonstrations of teaching activities, status reports on NASA projects and Space Shuttle launches, and more. The videoconference series is free to registered educational institutions. A C-band satellite receiving system is required (or arrangements must be made to receive the satellite signal through the local cable television system).

NASA's Central Operation of Resources for Educators (CORE)

NASA CORE
Lorain County JVS
15181 Route 58 South
Oberlin, OH 44074
Phone: (216) 774-1051, ext. 249/293
Fax: (216) 774-2144
E-mail: nasaco@leeca8.leeca.ohio.gov
Internet: spacelink.nsfc.nasa.gov/CORE

The Central Operation of Resources for Educators (CORE) is NASA's worldwide distribution center for multimedia educational materials. For educators who cannot access a Teacher Resource Center or Regional Teacher Resource Center, this mail order service is an excellent alternative that makes NASA material available to educators for a minimal fee. A free NASA CORE catalog is available upon request and includes such items as videotape programs, slide programs, computer software, NASA memorabilia, and other supplemental curriculum materials.

NASA
Teacher Resource Centers

NASA Ames Research Center
Teacher Resource Center
Mail Stop 253-2
Moffett Field, CA 94035-1000
Phone: (415) 604-3574

NASA Teacher Resource Center
for Dryden Flight Research Center
45108 N. 3rd Street East
Lancaster, CA 93535
Phone: (805) 948-7347

NASA Goddard
Space Flight Center
Teacher Resource Laboratory
Mail Code 130.3
Greenbelt, MD 20771-0001
Phone: (301) 286-8570
Internet: pao.gsfc.nasa.gov

NASA Goddard Space Flight Center
Wallops Flight Facility
Education Complex - Visitor Center
Teacher Resource Lab
Bldg. J-17
Wallops Island, VA 2337-5099
Phone: (804) 824-2297/2298

NASA Johnson Space Center
Teacher Resource Center
Mail Code AP2
2101 NASA Road One
Houston, TX 77058-3696
Phone: (281) 483-8696

NASA John F. Kennedy
Space Center
Educators Resources Laboratory
Mail Code ERL
Kennedy Space Center, FL 32899
Phone: (407) 867-4090

NASA Teacher Resource Center
for Langley Research Center at the
Virginia Air and Space Center
600 Settler's Landing Road
Hampton, VA 23669-4033
Phone: (757) 727-0900, ext. 757
Internet: seastar.vasc.nus.va.us

NASA Lewis Research Center
Teacher Resource Center
Mail Stop 8-1
21000 Brookpark Road
Cleveland, OH 44135-3191
Phone: (216) 433-2017

NASA Teacher Resource Center
for Marshall Space Flight Center
U.S. Space & Rocket Center
P.O. Box 070015
Huntsville, AL 35807-7015
Phone: (205) 544-5812

NASA Stennis Space Center
Teacher Resource Center
Building 1200
Stennis Space Center, MS 39529-6000
Phone: (601) 688-3338

NASA Jet Propulsion Laboratory
Teacher Resource Center
JPL Educational Outreach
4800 Oak Grove Drive
Mail Code CS-530
Pasadena, CA 91109-8099
Phone: (818) 354-6916

NASA Mobile Teacher Resource Center
NAFEO Services, Inc.
400 12th Street, N.E.
Washington, DC 20002

NASA's Mobile Teacher Resource Center (MTRC) is a self-contained resource facility designed to support teacher enhancement workshops in areas that do not have a Teacher Resource Center or Regional Teacher Resource Center located nearby. Contact NAFEO Services for additional information.

NASA
Regional Teacher Resource Centers (RTRC)

Alabama

Tri-State Learning Center
NASA Regional Teacher
Resource Center
P.O. Box 508
Iuka, MS 38852-0508
(601) 423-7455

Alaska

Alaska Science Center
NASA Regional Teacher
Resource Center
Alaska Pacific University
4101 University Drive
Anchorage, AK 99508
(907) 564-8207

Arizona

Lunar and Planetary Lab
NASA Regional Teacher
Resource Center
1629 E. University Boulevard
University of Arizona
Tucson, AZ 85721-0092
(520) 621-6939/6947
E-mail: lebofsky@u.arizona.edu

Arkansas

University of Arkansas - Little Rock
NASA Regional Teacher
Resource Center
Natural Science Bldg., Room 215
2801 South University
Little Rock, AR 72204
(501) 569-3259

California

NASA San Joaquin Valley Regional
Teacher Resource Center
California State University, Fresno
5005 N. Maple Avenue - Mail Stop 01
Fresno, CA 93740-0001
(209) 278-0355

Colorado

U.S. Space Foundation
NASA Regional Teacher
Resource Center
2860 S. Circle Drive, Suite 2301
Colorado Springs, CO 80906-4184
(719) 576-8000

Connecticut

Aerospace and Environmental Education Resource Center
Media Building, Room 138
Eastern Connecticut State College
83 Windham Street
Willimantic, CT 06226-2295
(203) 465-5725

Delaware

Delaware Aerospace Center
NASA Regional Teacher
Resource Center
500 C Duncan Road
Wilmington, DE 19809-2359
(302) 761-7494

District of Columbia

National Air and Space Museum
Smithsonian Institution
Education Resource Center, MRC-305
Washington, DC 20560
(202) 786-2109

University of the District of Columbia
NASA Regional
Teacher Resource Center
Mail Stop 4201
4200 Connecticut Avenue, N.W.
Washington, D.C. 20008
(202) 274-6287

Georgia

Southern College of Technology/ GYSTC
1100 S. Marietta Parkway
Marietta, GA 30060-2896
(770) 528-6272
E-mail: adocal@sct.edu

Hawaii

Barbers Point Elementary School
NASA Regional
Teacher Resource Center
Boxer Road
Barbers Point Naval Air Station
Ewa Beach, HI 96706
(808) 885-6030

Idaho

University of Idaho at Moscow
NASA Regional
Teacher Resource Center
ID Space Grant College
College of Education
Moscow, ID 83844-3080
(208) 885-6030

Illinois

Chicago Museum of
Science and Industry
NASA Regional
Teacher Resource Center
57th Street and Lakeshore Drive
Chicago, IL 60637-2093
(312) 684-1414, ext. 2426

Parks College of St. Louis University
NASA Regional
Teacher Resource Center
Rt. 157 and Falling Springs Road
Cahokia, IL 62206
(618) 337-7500

Indiana

University of Evansville
NASA Regional
Teacher Resource Center
School of Education
1800 Lincoln Avenue
Evansville, IN 47722
(812) 479-2393

Iowa

University of Northern Iowa
NASA Regional
Teacher Resource Center
IRTS, Room 222
Schindler Education Center
Cedar Falls, IA 50614-0009
(319) 273-6066

Kansas

Kansas Cosmosphere & Space Center
NASA Regional
Teacher Resource Center
1100 North Plum
Hutchinson, KS 67501-1499
(316) 662-2305

Kentucky

Murray State University
NASA Regional
Teacher Resource Center
Waterfield Library
Murray, KY 42071-0009
(502) 762-2850

Louisiana

Bossier Parish Community College
NASA Regional
Teacher Resource Center
2719 Airline Drive
Bossier City, LA 71111
(318) 746-7754

Southern University - Shreveport
NASA Regional
Teacher Resource Center
Downtown Metro Center
610 Texas Street
Shreveport, LA 71101
(318) 674-3444

Massachusetts

Bridgewater State College
NASA Regional
Teacher Resource Center
Maxwell Library
Media Service
Bridgewater, MA 02325
(508) 697-1242

Michigan

Central Michigan University
NASA Regional
Teacher Resource Center
Ronan Hall, 101
Mount Pleasant, MI 48859
(517) 744-4387

Northern Michigan University
The Glenn T. Seaborg Center
NASA Regional
Teacher Resource Center
1401 Presque Isle
Marquette, MI 49855-5394
(906) 227-2002
E-mail: seaborg@nmu.edu

Oakland University
NASA Regional
Teacher Resource Center
O'Dowd Hall, Room 216
Rochester, MI 48309-4401
(313) 370-2485/4230

Minnesota

Mankato State University
NASA Regional
Teacher Resource Center
Department of Curriculum
and Instruction
MSU Box 52/P.O. Box 8400
Mankato, MN 56002-8400
(507) 389-1516

Saint Cloud State University
Center for Information Media
NASA Regional
Teacher Resource Center
720 4th Avenue South
St. Cloud, MN 56301
(612) 255-2062

Mississippi

Mississippi Delta Community College
NASA Regional
Teacher Resource Center
P.O. Box 668
Moorhead, MS 38761
(601) 246-6383

Tri-State Learning Center
NASA Regional
Teacher Resource Center
P.O. Box 508
Iuka, MS 38852-0508
(601) 423-7455

Mississippi Band of Choctaw Indians
Teacher Enhancement Center
Route 7, Box 72
Philadelphia, MS 39350
(601) 656-9638

Montana

Western Montana College of the
University of Montana
NASA Regional
Teacher Resource Center
Carson Library
710 South Atlantic
Dillon, MT 59725
(406) 683-7541

Nebraska

University of Nebraska State Museum
NASA Regional
Teacher Resource Center
14th & U Streets
307 Morrill Hall
Lincoln, NE 68588-0338
(402) 472-6302

University of Nebraska at Omaha
NASA Regional
Teacher Resource Center
Mallory Kountze Planetarium
Durham Science Center, Room 144
60th and Dodge Street
Omaha, NE 68182-0266
(402) 554-2510

Nevada

NASA/Nevada Regional
Teacher Resource Center - C2A
Community College of
Southern Nevada
3200 E. Cheyenne Avenue
North Las Vegas, NV 89030-4296
(702) 651-4505

New Jersey

Georgian Court College
NASA Regional
Teacher Resource Center
900 Lakewood Avenue
Lakewood, NJ 08701-2697
(908) 364-2200, ext. 479

New Mexico

New Mexico State University
NASA Regional
Teacher Resource Center
New Mexico Space Grant Consortium
Box 30001, Dept. SG
Las Cruces, NM 88003-0001
(505) 646-6414

University of New Mexico
NASA Regional Teacher Resource
Center
Continuing Education and
Community Service
1634 University Blvd., N.E.
Albuquerque, NM 87131-4006
(505) 277-3861

New York

The City College
NASA Regional
Teacher Resource Center
Harris Hall, Room 109
Convent Avenue at 138th Street
New York, NY 10031
(212) 650-6993

North Carolina

University of North Carolina - Charlotte
NASA Regional
Teacher Resource Center
J. Murray Atkins Library
Charlotte, NC 28223
(704) 547-2559
E-mail: nasartrc@unccvm.uncc.edu

North Dakota

University of North Dakota
NASA Regional
Teacher Resource Center
The Wayne Peterson Room
Clifford Hall - 5th Floor
Space Studies Department
P.O. Box 9008, University Station
Grand Forks, ND 58202-9008
(701) 777-4856

Ohio

University of Cincinnati
NASA Regional
Teacher Resource Center
Curriculum Resources Center Library
Mail Location 0219
600 Blegen Library
Cincinnati, OH 45221-0219
(513) 556-1430

Oklahoma

Oklahoma State University
O.S.U. Aerospace Professional
Development Center
308 A CITD
Stillwater, OK 74078-8089
(405) 744-6784
(405) 744-6785 Fax

Oregon

Oregon Museum of
Science and Industry
NASA Regional
Teacher Resource Center
Science Program Department
1945 SE Water Avenue
Portland, OR 97214-3354
(503) 797-4579

Pennsylvania

University of Pittsburgh
School of Education
Computer & Curriculum Inquiry Center
NASA Regional
Teacher Resource Center
230 S. Boquet Street
Pittsburgh, PA 15260
(412) 648-7560/7558

Puerto Rico

University of Puerto Rico at Mayaguez
Resource Center for
Science and Engineering
NASA Regional
Teacher Resource Center
Physics Building #200
Mayaguez, PR 00681
(787) 831-1022/1025
(787) 832-4680 Fax
E-mail:
    m_schwarz@rumac.upr.clu.edu

Rhode Island

Rhode Island College
NASA Regional
Teacher Resource Center
Henry Barnard Elementary School
600 Mt. Pleasant Avenue
Providence, RI 02908
(401) 456-8001

South Carolina

Stanback Planetarium
NASA Regional
Teacher Resource Center
P.O. Box 7636
South Carolina State University
Orangeburg, SC 29117-7636
(803) 536-8709/8711/7174
E-mail: starman@scsv.edu

South Dakota

NASA Regional
Teacher Resource Center
T.I.E. Office
1925 Plaza Blvd.
Rapid City, SD 57702
(605) 394-1876

Tennessee

Tri-State Learning Center
NASA Regional
Teacher Resource Center
P.O. Box 508
Iuka, MS 38852-0508
(601) 423-7455

University of Tennessee at Martin
NASA Regional
Teacher Resource Center
Center for Excellence in
Mathematics and Science
Martin, TN 38238-5029
(901) 587-7191/7166

Texas

University of Texas at Brownsville
NASA Regional
Teacher Resource Center
80 Fort Brown
Brownsville, TX 78520
(210) 982-0295
(210) 544-5495
E-mail: dnutter@utb.edu

Utah

Utah State University
NASA Regional
Teacher Resource Center
Educational Resources and
Technology Center
Logan, UT 84322-2845
(801) 797-3377
Internet: www.teacherlink.usu.edu/
  nasa.index.html

Weber State University
NASA Regional
Teacher Resource Center
Curriculum Library
College of Education
Ogden, UT 84408-1302
(801) 626-7614/6279

Vermont

Norwich University
Vermont College
Educational Resource Center
NASA Regional
Teacher Resource Center
Schulmaier Hall
Montpelier, VT 05602
(802) 828-8845

Virginia

Radford University
NASA Regional
Teacher Resource Center
P.O. Box 6999 Walker Hall
Radford, VA 24142
(540) 831-6284

Washington

University of Washington
NASA Regional
Teacher Resource Center
AK-50, c/o Geophysics Department
Seattle, WA 98195
(206) 543-1943

West Virginia

Wheeling Jesuit College
NASA Regional
Teacher Resource Center
316 Washington Avenue
Wheeling, WV 26003
(304) 243-2388

University of Charleston
NASA Regional
Teacher Resource Center
2300 MacCorkle Avenue SE
Charleston, WV 25304-1009
(304) 357-4707

Wisconsin

University of Wisconsin at LaCrosse
NASA Regional
Teacher Resource Center
Morris Hall, Room 200
LaCrosse, WI 54601
(608) 785-8148/8650

Wyoming

University of Wyoming
NASA Regional
Teacher Resource Center
Learning Resource Center
P.O. Box 3374 University Station
Laramie, WY 82071-3374
(307) 766-2527

Appendix D
NASA Sites and Information

The following information was extracted from the NASA Web site (http://www.nasa.gov) which contains a wide variety of information on NASA , its programs and its publications.

NASA Information Center

NASA Headquarters Information Center
Mail Code CMI-1
300 E Street SW
Room 1H23
Washington, D.C. 20546-0001
Phone: (202) 358-0000

The NASA Information Center serves NASA Headquarters and NASA employees as well as government contractors and officials, educators, and members of the national and international public. Inquiries are received in person, by phone, by mail and by electronic mail.

The Information Center staff provide directional and core information concerning the functions, activities and personnel of the Agency to approximately 7,000 persons who call, visit, or write NASA Headquarters each month. The staff also respond to requests for publications and reports produced by NASA. The Center hours are 7:30 AM to 4:30 PM, Monday through Friday.

NASA Points of Contact

Headquarters
National Aeronautics and Space
Administration (NASA)
Washington DC 20546-0001
Phone: (202) 358-0000

Ames Research Center (ARC)
National Aeronautics and Space Administration (NASA)
Moffett Field, CA 94035-1000
Phone: (415) 604-5000

Dryden Flight Research Center (DFRC)
NASA
P.O. Box 273
Edwards, CA 93523-0273
Phone: (805) 258-3311

KSC VLS Resident Office
(Vandenberg AFB)
NASA
P.O. Box 425
Lompoc, CA 93438
Phone: (805) 866-5859

Goddard Space Flight Center (GSFC)
NASA
Greenbelt Road
Greenbelt, MD 20771-0001
Phone: (301) 286-2000

Langley Research Center (LaRC)
NASA
Hampton, VA 23681-0001
Phone: (804) 864-1000

Goddard Institute for Space Studies
Goddard Space Flight Center
NASA
2880 Broadway
New York, NY 10025
Phone: (212) 678-5500

Lewis Research Center (LeRC)
NASA
21000 Brookpark Road
Cleveland, OH 44135-3191
Phone: (216) 433-4000

Jet Propulsion Laboratory (JPL)
NASA
4800 Oak Grove Drive
Pasadena, CA 91109-8099
Phone: (818) 354-4321

George C. Marshall Space Flight Center
NASA
Marshall Space Flight Center, AL
35812-0001
Phone: (205) 544-2121

Lyndon B. Johnson Space Center (JSC)
NASA
Houston, TX 77058-3696
Phone: (713) 483-0123

Michoud Assembly Facility (MAF)
NASA
P.O. Box 29300
New Orleans, LA 70189
Phone: (504) 257-3311

John F. Kennedy Space Center (KSC)
NASA
Kennedy Space Center, FL 32899-0001
Phone: (407) 867-7110

NMO-JPL: NASA Management Office-
JPL
NASA
4800 Oak Grove Drive
Pasadena, CA 91109
Phone: (818) 354-5359

Slidell Computer Complex (SCC)
1010 Gause Boulevard
Slidell, VA 74058
Phone: (504) 646-7200

John C. Stennis Space Center (SSC)
NASA
Stennis Space Center, MS 39529-6000
Phone: (601) 688-2211

Wallops Flight Facility (WFF)
Goddard Space Flight Center
NASA
Wallops Island, VA 23337-5099
Phone: (804) 824-1000

White Sands Test Facility (WSTF)
Johnson Space Center
NASA
P.O. Drawer NM
Las Cruces, NM 88004-0020
Phone: (505) 524-5771

Overseas Points of Contact

Moscow

NASA CIS Representative
US Embassy-Moscow
PSC 77/NASA APO AE 09721
Phone: 9-7095-956-4271

Spain (Tracking)

ATTN: NASA Representative
PSC #61, Box 0037
APO AE 09642

European

NASA European Representative
US Embassy-Paris
PSC 116/NASA APO AE 09777-9200
Phone: 9-1-33-1-4265-8762

Australia (Tracking)

ATTN: NASA Representative
APO AP 96549

Freedom of Information Act

Freedom of Information Act (FOIA) requests should be sent to:

Patricia Riep-Dice
Mail Code PSN
NASA Headquarters
Washington DC 20546-0001
Phone: (202) 358-1764

Astronaut Information:

Those wanting information on NASA astronauts or related information can contact the Astronaut Office at:

Astronaut Office/CB
NASA
Johnson Space Center
Houston, TX 77058

Astronaut Addresses:

A few of NASA's former astronauts are listed below. When writing to them, in order to receive a response, please include a self-addressed, self- stamped envelope.

Dr. Buzz Aldrin
Starcraft Enterprise
233 Emerald Bay
Laguna Beach, CA 92651

Neil A. Armstrong
P.O. Box 436
Lebanon, OH 45036

Dr. Mae Jemison
The Jemison Group
PO Box 591455
Houston, TX 77259-1455

James A Lovell
President, Lovell Communications
PO Box 49
Lake Forest, IL 60045

Dr. Sally K Ride
Director, California Space Institute
University of California at San Diego
La Jolla, CA 92093

Master Index